DATE DUE

APPLES

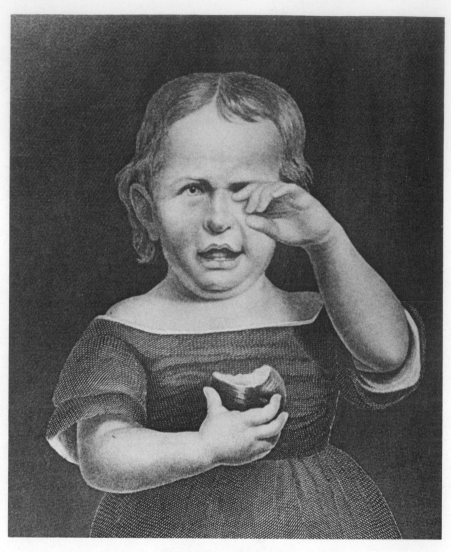

WHO'S BIT MY APPLE?
An engraving from *Peterson's Magazine*, September 1865.

APPLES

Peter Wynne

HAWTHORN BOOKS, INC.
PUBLISHERS / *New York*

To Lorry
ἡ καλὴ λαβέτω

APPLES

Copyright © 1975 by Peter Wynne. Copyright under International and Pan-American Copyright Conventions. All rights reserved, including the right to reproduce this book or portions thereof in any form, except for the inclusion of brief quotations in a review. All inquiries should be addressed to Hawthorn Books, Inc., 260 Madison Avenue, New York, New York 10016. This book was manufactured in the United States of America and published simultaneously in Canada by Prentice-Hall of Canada, Limited, 1870 Birchmount Road, Scarborough, Ontario.

Library of Congress Catalog Card Number: 74–15646

ISBN: 0–8015–0340–X

1 2 3 4 5 6 7 8 9 10

Recipe for "Rissoles on a Fish Day" reprinted from *Le Menagier de Paris* (circa 1393) translated by Eileen Power as *The Goodman of Paris*, London, 1928, with permission of Routledge & Kegan Paul Ltd., London.

Contents

PART II: RECIPES

Acknowledgments

One cannot hope to be entirely original when preparing a book about a topic as ancient and far-ranging as the apple. Much of this book is based on the works of others. Hopefully, my contribution has been to gather the data into one place and to interpret and organize some of the more difficult material.

Much of the historical material was culled from three sources: A. J. Downing's *The Fruit and Fruit Trees of America*, S. A. Beach's *The Apples of New York State*, and U. P. Hedrick's *A History of Horticulture in America to 1860*.

The folklore chapters center around the notion that once upon a time there was an apple goddess, an idea partly inspired by Robert Graves's *The White Goddess*. Much of the material on ancient apple lore was taken from B. O. Foster's essay "Notes on the Symbolism of the Apple in Classical Antiquity," a marvelous piece of scholarship but one not easily accessible since all the illustrative material is quoted in ancient Greek and Latin. The translations from Greek and Latin in this book are my own, and if there are any errors, they are mine, too. Basic to the idea of the apple goddess are the concepts of the sacred king and "dying and reviving" god, the brainchildren of Sir James George Frazer in his *The Golden Bough*.

The chapter on growing apples was drawn from two sources; a year of practical experience working for a tree surgeon and the scores of pamphlets issued by the New Jersey and United States departments of agriculture. The recipes and history are the joint products of ten years of cooking experience and three years of writing articles on food and cooking history for a newspaper, with all of the research and reading that those two activities entailed.

ACKNOWLEDGMENTS

On a more personal level, I must acknowledge a debt to several people who helped me to assemble the material for this book. Thanks must go to Maureen Taffe, principal librarian of the Johnson Public Library of Hackensack, New Jersey, and to her colleagues, who helped track down and obtain out-of-print books; to Alice Lydecker, who graciously gave me access to her personal library; to Elizabeth Cornelia Hall, curator of rare books at the Horticultural Society of New York; to Adele Zydel, who helped transform a hash of notes and worksheets into printable recipes; to Milisa Vlaisavljevic and Helen Shedd, who copied recipes from Samuel Pegge's *The Forme of Cury*, a book that seems to be missing from most American libraries that ever owned a copy; and finally to Sandra Choron, who blue-penciled the manuscript to sanity.

Part I

APPLES

A NEW
Orchard and Garden,

Or

T he beſt way for planting, grafting, and to make *any ground good, for a rich Orchard: Particularly in the North,* and generally for the whole kingdome of England, as in nature, *reaſon, ſituation, and all probability, may and doth appeare.*

With the Country Houſewifes Garden for hearbes of common uſe, their vertues, ſeaſons, profits, ornaments, variety of knots, models for trees, and plots for the beſt ordering of Grounds and Walkes.

As alſo the Husbandry of Bees, with their ſeverall uſes and annoyances, all being the experience of 48. yeares labour, and now the third time corrected and much enlarged, by *William Lawſon.*

Whereunto is newly added the Art of propagating Plants, with the true ordering of all manner of Fruits, in their gathering, carrying home, and preſervation.

Skill and paines bring fruitfull gaines.

Nemoſibinatus.

London, Printed by *Edward Griffin* for *Iohn Harison,* at the **golden** Vnicorne in Pater noſter-row. 1638.

This seventeenth-century illustration shows various orchard operations. At left, pruning; at center, planting; at right, setting a cutting so it will take root in the earth. From William Lawson's *A New Orchard and Garden*, London, 1638. (*Courtesy the Library of the New York Botanical Garden*)

1

The History of the Apple

The story of the apple began many centuries before man knew how to record the details, but it is more or less possible to reconstruct the events.

There is fossil evidence that the crab apples growing in Europe today grew there in prehistoric times. Like the modern crab tree, the prehistoric ones bore fruit that weighed about half an ounce. Then something happened to cause the size of the fruit to swell. Cultivation by early man may have been a factor, but the likeliest explanation is that two different species of crab apples cross-pollinated, and for some genetic reason the tree produced by this mismatch began to bear fruit many times larger than the fruit of either of its parents. The cross-breeding probably occurred many times, back and forth, but however it happened, a new strain of tree was formed, a tree that bore fruit eight or more times greater than a wild crab apple.

By the time recorded history began, the modern apple tree was a reality, and man soon learned (if he didn't know already) how to propagate and improve the varieties of apples that by then grew throughout most of Europe. But that is where the history of the apple begins.

The Ancient Apple

The bulk of scientific evidence indicates that the first true apples grew in the Caucasus and that they were the result of cross-breeding between Asiatic and European crab apples, which are both found today in their original forms in that region. Just as important, the Caucasus was home to a tribe of humans—the Caucasians or Aryans or Indo-Europeans, whatever you might want to call them. This tribe was the ancestor of most of the peoples of modern Europe, of the Persians, of the Afghans, and of many of the inhabitants of India. And it seems that wherever these Caucasians went they took the apple with them.

Of course, fruits of the apple family were known in Europe and the Middle East before the arrival of the Aryan peoples. Carbonized apples have been found among the remains of prehistoric, non-Aryan tribes living in Switzerland. Furthermore, some of these fossil apples were bigger than typical wild European crab apples, probably as a result of cultivation. At the same time, these ancient Swiss apples seem to have been produced by trees of the species *Malus sylvestris*, whereas the fruits we know as apples are produced by the Aryan species *Pyrus malus*.

Tracing the very earliest history of the apple is no easy matter. Homer, who lived sometime between the twelfth and seventh centuries B.C., provides the earliest mention of the fruit. He relates in *The Odyssey* that apple trees grew in the gardens of King Alcinous and Laertes, father of Odysseus. Unhappily, the poet did not provide a detailed description of the trees or their fruit, and he used the word *melon*, which sometimes was used as a generic term for any fruit that grew on a tree.

Fortunately, later Greek writers were usually quite careful to distinguish the apples the Greek tribes had known from the earliest times from the members of the apple family the Greeks had first encountered when they migrated into the Mediterranean area. The quince, for example, was known in the eastern Mediterranean long before the Aryans arrived, and this fruit doubtlessly was the "apple" of the biblical Song of Solomon. The later writers call the quince a Cretan apple, after the people of the island of Crete, who apparently introduced the Greeks to that fruit.

Actually, there is very little information about apples and apple-growing among the Greeks and Romans until quite late in their respective histories. It is likely that the apple was an important item in the diet of the Aryan tribes when they were in the Caucasus, a cool, temperate region where apples thrive. However, when these tribes moved down to the warmer Mediterranean region, they adopted the agricultural traditions of the peoples already living there, whose diets centered around meats and grains; such vegetables as onions, leeks, cucumbers, and squashes; and such fruits as grapes, figs, and olives. The apple continued to play a considerable role in the myths and legends of the Greeks and Romans, but it did not have a major part in their respective diets once they reached the Mediterranean.

The later Greeks and Romans prized the apple highly, regarding it as something of a luxury item. The Greek historian Plutarch once wrote in his essay "The Symposiacs" of an imaginary banquet at which the table talk turned to the apple. The question was posed, Why is the apple tree called the bearer of splendid fruit? One of the banqueters replied that the best qualities of all fruits were combined in the apple, that the apple was smooth to the touch, that it imparted a sweet

This nineteenth-century engraving shows two carbonized crab apples, fossils of a sort, found among the remains of prehistoric lake dwellers in Switzerland. Caches of these tiny apples, some with as many as three hundred fruits, have been found at several Swiss locations. About an inch in diameter, these specimens have been identified as belonging to the same species of crab apple that grows wild in Switzerland today. Most of the fruits had been cut into halves, thirds, or quarters, apparently to dry for winter use. Interestingly, several of the apples show evidence of codling moth damage, and that pest is still active today. From *Harper's New Monthly Magazine*, August 1875.

odor to the hand without soiling it, that the apple was sweet to the taste and pleasing to eyes and nose. And so, the guest concluded, the apple tree should be called "the bearer of splendid fruit," because of all fruits only the apple could please all the senses at once.

The Romans were also capable of waxing poetic about the apple, but of much greater interest is the legacy of farm literature they left to the world. Marcus Portius Cato (the elder), who lived in the second century B.C., authored a remarkable work on farming called *De Agricultura*, a title as simple and direct as the book itself. (The Romans of the old republic could be a stern lot, and Cato was one of the sternest.) At first glance, Cato's book seems impossibly primitive, detailing sacrifices to tree spirits and providing hints on the treatment of slaves, but much of his advice on the care and keeping of plants is as valid today as it was more than two thousand years ago.

As a method of grafting twigs to larger branches, Cato suggests cutting off the end of the stock, the branch on which the graft would be placed, on a slant so that water would not collect at the junction point and driving a sharpened hardwood stick between the bark and wood of the stock branch and placing the twig to be grafted, sharpened similarly, into the space made by the hardwood stick. He cautions the farmer not to tear the bark of the branch: the inner bark of the twig must make contact with the inner bark of the branch for the graft to take. This method is practically identical with the modern method of cleft grafting, in which a pair of sharpened twigs are inserted into a split in the end of the stock branch.

Cato then suggests covering the graft with a kind of plaster made of clay, sand, and dung. This may sound horribly primitive, yet in nineteenth-century America the same mixture, essentially, was still in use. Andrew Jackson Downing's very respected *The Fruits and Fruit Trees of America*, first published in 1845, contains this recipe: "Grafting clay is prepared by mixing one third horse-dung free from straw, and two thirds clay, or clayey loam, with a little hair, like that used in plaster, to prevent its cracking."

Cato also mentions a type of bud grafting in which a bud-bearing piece of bark is fitted into an opening cut in the stock plant. He also details methods of propagating fruit trees by taking cut branches and putting them in the earth so that they grow roots; by bending down and burying branches that are still attached to the tree, a technique known today as layering; and finally, by a sophisticated method called air layering.

Cato suggests pushing a small, slender branch still attached to its tree through a hole in the bottom of a basket. The basket should then be tied in place, filled with earth, and left among the upper branches of the tree. Two years later the branch should be cut beneath the basket, and the new tree and basket should be planted. This method is still used today to propagate certain plants that grow poorly from seed.

This fact that some plants grow poorly from seed is essential to understanding why Cato, who was a practical man and would not have wasted time with sophistication for its own sake, lavished so much attention on grafting and vegetative methods of propagation.

Apple trees grow from seeds well enough. However, there is no guarantee of what kind of fruit a seed-grown tree will produce. Like a human, an apple tree grown from seed has two parents. Also like a human such a tree may or may not resemble its parents. Actually, the odds are that an apple tree grown from seed will produce very small, sour fruit, as if the tree has "returned to the wild." Only rarely will the seed-grown tree be superior to either parent. But an apple tree propagated vegetatively, through cuttings or layering, will produce exactly the same fruit that its single parent produced.

The fact that an offspring from a sexually mated pair will very likely inherit the worst qualities of both begins to sound as though Mother Nature is cruel and arbitrary, but that is far from the case. What we consider good eating or cooking apples are actually freaks—freakishly big, freakishly sweet, growing on trees that are freakishly unproductive in terms of quantity. Nature prefers a tree that produces thousands of sour little apples in the hope that more fruit will be left uneaten, thus giving the tree a greater chance of reproducing itself.

The Romans didn't have our modern, genetic explanation for why apple trees grown from seed "returned to the wild," but these crafty farmers certainly knew how to get around the problem. Further, if they happened to have a strong tree that produced inferior fruit, they could get around that, too. They could graft buds and branches from another, more desirable-fruiting tree onto it. When the grafted branches grew out, they would produce the same superior fruit that their parent had produced, for they had the same genetic structure as the parent tree.

Actually, Cato gave short shrift to the apple. He mentions only pears and other "flesh fruits." However, any technique of grafting or propagation appropriate to the pear can be used successfully with the apple. But to find out what Roman apples were actually like, we have to rely on another author—Pliny.

Pliny (his full name was Caius Plinius Secundus) wrote a huge work entitled *Historia Naturalis* ("Natural History"), which he completed in A.D. 77. The work consists of thirty-seven volumes, and scattered among its pages is a rather good description of the apples of antiquity. One sometimes encounters the claim that Pliny described thirty-six varieties of apples, but this is entirely inaccurate. Many of his "apples" were actually quinces, medlars (Eurasian crab-applelike fruits), and who knows what else. Pliny classifies citrons, peaches, even figs, as apples, but some of the fruits he mentions were quite likely apples we would recognize. He writes of a Syrian red apple, named for its color, and of an apple called orthomastion, meaning "upturned breast" in Greek, because it was thought to resemble a woman's bosom. He also writes of a "must-apple," so named because it ripens early, but called the honey apple in his day for its flavor.

Pliny's work is full of peculiar notions. For example, he writes that blood-red apples are the result of apples having been grafted onto mulberry trees, a graft that in actuality just would not be able to "take" and even if it could, wouldn't affect the color of the fruit. Pliny offers the fanciful notion of the mulberry–apple graft to explain why some apples are completely red. He also says that all apples have red on the cheek exposed to sunlight. This information, added to the fact that apples were called golden in classical literature, leads one to suspect that yellow apples with red cheeks were by far the most common sort in ancient Greece and Rome.

Oddly, one of the few apples grown today that may be a descendant of the apples of Rome is golden yellow and has a red cheek when ripe.

Apy

The Apy, or Pomme d'Api, is known in the English-speaking world as the Lady Apple. A small winter dessert apple, Apy is said to be the same as the Malum Appianum of ancient Rome. This eighteenth-century bookplate shows a cluster of immature fruit, blossoms, seeds, and mature fruit—whole and in cross section. From Henri Louis Duhamel Du Monceau's *Traité des Arbres Fruitiers*, Paris, 1782. (*Courtesy the Library of the New York Botanical Garden*)

We know this flavorful little apple as the Lady Apple. In France, it is called *pomme d'Api*, which means "Appian apple." Pliny also mentions an apple by this name. He says this fruit was developed by a man named Appius, who grafted a quince onto an apple. As a result of this graft, the apple had the flavor of a quince, the size of an apple, and a skin with a rosy blush, which seems to fit the description of our Lady Apple.

Pliny's theory about the graft is apocryphal; actually, as explained above, the stock plant has no effect on the type of fruit the grafted branch produces. Another, more plausible story maintains that Appius Claudius Caeca brought his namesake apple from Greece to Rome sometime in the fourth century B.C. and that the Romans planted this apple throughout their empire in later centuries.

Of course, proving that the Lady Apple is truly the Appian apple of ancient Rome and not a later variety is out of the question. Twenty-five-hundred years is a long time, and the record book just isn't complete enough. However, it is pleasant to believe that a fruit grown today could have such a long and wonderful lineage.

Finally, we cannot leave the ancients without mentioning some of the hints Pliny provides on when and how to pick apples and how to store them. He suggests picking apples after the autumnal equinox, no earlier than the sixteenth and no later than the twenty-eighth day after the new moon. Most years, this would be from the last week of September through the first week of October. Because excessive moisture hastens decay, he cautions against picking on rainy days or until an hour after sunrise (when the dew would have dried). This pretty much conforms with standard practice right into this century. Pliny also cautions against mixing windfallen apples with those that have been picked. Windfalls, of course, are subject to bruising and subsequent deterioration.

He advises storing apples in a cool, dry room with windows on the north side that can be opened on sunny days. The floor should be planked and covered with a layer of close-packed straw or with the chaff from grain. The apples should be laid out in rows, with spaces in between to allow for air circulation. Until cold storage became practical in the twentieth century, no one had come up with a substantially better system than Pliny's, at least none that could be practiced easily on a large scale.

Pliny also mentions some alternative ways of storing apples. Most people, he says, put apples into a two-foot-deep hole, with a layer of sand beneath them, then cover the hole with an earthenware lid, which they then cover with earth. Some of the finest apples were smeared with plaster or wax. Some people put apples in clay pots, one to a pot, sealed the pots with pitch, then put them into a barrel. Others packed the apples in wool inside a wooden case, which then was sealed with a mixture of wet clay and chaff. At least one of these methods—storage underground in a sand-lined pit—is in use on farms today.

Pliny, incidentally, also describes the various modes of grafting and propagating fruit trees vegetatively. He recommends growing apple trees from seed to provide strong stocks that can be improved by grafting. Like Cato, he was thoroughly familiar with the use of manures and other fertilizers and even with the use of "green manures." These are plants, such as the vetches and other members of the pea family, that have bacteria in their roots that can remove nitrogen from the air to enrich the soil.

Like Cato, Pliny also mentions one of the early methods of insect control. In one passage he recommends smearing the trunks and lower branches of vines and trees with a sticky substance to catch caterpillars and other insects that hatch in the soil, then climb up plants to feed on their leaves. Then, almost as if to reassert that he was really an ancient, he mentions another way of fighting caterpillars: Have a woman who has just begun to menstruate walk around each tree barefoot and with her belt untied.

Although both Cato and Pliny include several types of recipes in their respective works, neither author details one for apples, and the only such recipe that survives from ancient Rome comes from a book called *De Re Coquinaria* ("On Cookery"), which dates from about the third century but is attributed to a gourmand named Apicius, who lived two centuries before. The dish is called *Minutal Matianus* ("Diced Pork and Matian Apples"), and the recipe can be found on page 180.

Pliny died in A.D. 79, two years after completing his *Historia*

Naturalis, and just about four centuries later (A.D. 476) Rome joined him. In that year, a "barbarian" named Odoacer was proclaimed the ruler of all of Italy. However, changes in the social structure of Rome, which may or may not have contributed to its downfall, had already been well under way in Cato's day, in the early second century B.C. Cato railed against luxury and affirmed in his book his belief in the value of good hard work.

Pliny, three centuries later, was decrying the decline of Roman agriculture and blaming it on the almost exclusive use of slave labor in farming. Earth was delighted, Pliny writes, when centuries before it had been tilled with the laurel-decked plowshares of the conquering generals of Rome. He hints at the underlying reason: When the land-owner did his own work, it was done with greater care and attention.

Whether the Roman slaves ever knew very much about fruit culture is moot, but indications are that they didn't. And when the noble classes of Rome forsook their farms for lives of luxury in the teeming capital, perhaps they forgot the centuries of farm lore that was their inheritance.

Christianity slowly spread throughout the Roman Empire, and perhaps it was that religion's tendency to regard life beyond the grave as more important than life on earth that caused stagnation in the sphere of fruit culture. The pagan Romans had been a very worldly people, and their practical, workaday attitude permeated all phases of their lives. They developed and practiced some of the most sophisticated techniques of agriculture seen to that time. The same cannot be said for the Christians who inherited the Roman world.

The Apple in Europe

It seems that Europe largely ignored the systematic, almost scientific principles of agriculture that the Romans had devised. Crop rotation and the use of green manures, which the Romans had known about, were not even common knowledge in nineteenth-century America. Europe seems to have relearned the art of fruit growing from the followers of Islam, who had conquered most of Spain and parts of southern France by the eighth century.

Between A.D. 500 to 1500 cookery made some progress, with the development of dishes quite unlike those prepared in ancient Rome. Several apple recipes have come down to us from the Middle Ages. A kind of fried patty made of apples, nuts, and rice was included in a book called *Le Managier de Paris* ("The Householder of Paris"), which was written about 1393 by an elderly gentleman as a book on housekeeping for his young bride. The recipe for Rissoles on a Fish Day can be found on page 193.

Other dishes, such as Appulmose (page 201), a kind of applesauce and even a recipe for Apple Blossom Caudle (page 207), were included in a cookbook compiled in 1390 by the chefs of King Richard II. The collection was edited and published in 1780 by the scholar and antiquarian Samuel Pegge under the title *The Forme of Cury* ("cookery").

During the fifteenth century the art of grafting fruit trees may not have been entirely ignored. A fifteenth-century manuscript survives today which mentions grafting of apple trees: "Iff thou wilte that thy appyllys be rede, take a graff of an appyltre and ympe [bind] hit opone a stoke of an elme or an eldre, and hit shalbe rede appylles." This bit of misadvice is preserved in a book called *Early English Miscellanies in Prose and Verse*, published in 1855. It shows the same lack of understanding of the principles of grafting that Pliny revealed when he suggested that all-red apples could be produced by grafting apple branches to mulberry trees. One gets the impression that each generation of scholars passed this kind of misinformation on to the next but that none of them actually attempted to grow or graft an apple tree.

It is clear that by the late 1500s, knowledge of the niceties of fruit growing was at least in the hands of British aristocrats. In *The Herball, or Generall Historie of Plantes*, published in 1596, Jean Gerarde urged the British gentry to propagate only the best grafted trees on their estates.

> The tame and graffed Apple trees are planted and set in gardens and orchards for that purpose: they delight to grow in good and fertill grounds. . . .

But I have seene in the pastures and hedge rowes about the grounds of a worshipfull Gentleman . . . so many trees of all sortes, that the servants drinke for the most part no other drinke, but that which is made of apples. The quantitie is such, that by the report of the Gentleman himselfe, the Parson hath for tithe many hogsheads of Syder. The hogs are fed with the fallings of them (the trees), which are so many, that they make choise of those Apples that they do eate, who will not taste of any but the best. An example doubtlesse to be followed of Gentlemen that have land and living: (but envie faith, the poore will breake downe our hedges, and we shall have the least part of the fruit) but forward in the name of God, graffe, set, plant and nourish up trees in every corner of your grounds, the labour is small, the cost is nothing, the commoditie is great, your selves shall have plentie, the poor shall have somewhat in time of want to relieve their necessitie, and God shall reward your good mindes and diligence.

This seventeenth-century bookplate shows three steps in the process of cleft grafting. At center, the orchardist sets a scion branch into a stock tree. At left is a finished graft; at right, a graft ready to face the weather. From *A Booke of the Arte and Maner Howe to Plante and Graffe*, translated from the Dutch by Leonard Mascall, London, 1572. (*Courtesy the Library of the New York Botanical Garden*)

*Dell'innestare gli Arbori con troncargli per
traucrſo, & fendergli per il lungo.*
Cap. XXXVIII.

Grafting practices in the seventeenth century were similar to those of today. The second figure from the left is a scion branch sharpened for insertion in a cleft cut into the stock tree. Second from right is the scion set in place. Other figures show cleft grafts bound and finally covered with straw to protect them from the elements. Modern practice would omit the straw covering, protecting the new graft from the weather with a mixture of rosin and wax. From Marco Bussato's *Giardino D'Agricultura*, Venice, 1612. (*Courtesy the Library of the New York Botanical Garden*)

From Gerarde's proselytizing it seems clear that while grafting was known, planting orchards exclusively with grafted trees was rare, even among the British aristocrats of his day. Actually, Gerarde hints at one of the explanations for why English apple growing (later in America as well) lagged behind that of the rest of Europe.

In England cider was an important beverage, whereas on the Continent, with a few exceptions, it was less so. Cider apples need not be of the same quality as dessert apples. In fact, the best cider comes from apples produced by seed-grown trees. Even today, good cider is pressed from a mixture of several different kinds of apples. This gives the drink a better balance—sweetness from one kind, tartness from another, fragrance from yet one more.

In an orchard planted with ungrafted, seed-grown trees, there are as many varieties as there are trees, because each tree has its own

Di Marco Buſſato. 22

fuora del terreno, ilqual terreno non ſia a guiſa di letame ca-
lido ; perche le palmuccie ſi rebolliriano a riſcaldandoſi, &
non ſariano buone da inneſtare , & eſſendo il terreno dolce
negro, potete ſepelire tutte le palmuccie, ma lodo che le
reſpiri.

*Del prendere dall'arbore le palmuccie , quando le comenẐano a
ingroſſar gli occhi per far le frondi, & conſeruar le
buone da inneſtare.* Cap. XXXV.

Pruning the apple tree in spring (a custom frowned on in the United
States). From Marco Bussato's *Giardino D'Agricultura*, Venice, 1612. (*Courtesy
the Library of the New York Botanical Garden*)

genetic makeup. Therefore, when making cider with apples taken from a seedling orchard, there is no need to mix varieties. The mixture is already there.

Beyond that, in a good-sized plantation of seedling trees, one is likely to find only one or two trees that produce fruit good enough for the table. In short, there wasn't that much pressure on the British gentry to improve the apples they grew, and the common people apparently didn't know how.

On the Continent, however, greater emphasis was placed on growing apples for dessert use, and very sophisticated methods of cultivation were devised in the process. The French, for example, perfected the

CIDER-MAKING IN NORMANDY

This nineteenth-century magazine illustration depicts the preparation of pomace, the ground-up apples, which then were placed into a cider press. The cider-maker shoveled the apples into the circular trough, where they were mashed by the huge, horse-drawn stone wheel shown here in the background. The wheel was on an axle, which in turn was fastened to a movable pivot (the post just behind the cider-maker). The horse was led around the outside of the trough, dragging the wheel, its harness attached to the wheel's axle. From the *Illustrated London News*, October 25, 1873.

technique of training apple trees to trellises. The trellises were called *espaliers*, and we call such trained trees by that same name. By pruning away every nonessential branch and supporting the tree by artificial means, French gardeners were able to induce it to bear fruit of the most extravagant size and quality.

The French also developed the dwarf apple tree by grafting apple branches onto the trunk of a native variety of wild apple, sometimes identified as a thorn tree. The effect of the graft was to stunt the growth of the apple branches but not of the fruit. The resulting tree was less than half the size of a standard apple tree, yet bore a large crop of full-sized fruit. Often the fruit was superior in quality to that produced by the full-sized "parent." The practical advantages of dwarf trees are more fully discussed in Chapter 3.

The knowledge of espaliers and dwarfs spread across Europe, but in England the real impetus toward better apple growing came with the Huguenots, who fled France in droves after the Edict of Nantes was rescinded in 1685. That edict, published by Henri IV in 1598, had provided guarantees of religious freedom to the Protestant Huguenots. When Louis XIV rescinded the edict, no fewer than 40,000 Huguenots, many of them skilled artisans, fled to England, Holland, Switzerland, and America, bringing with them knowledge of the more sophisticated methods of apple cultivation.

The American Apple

The history of the apple in America paralleled that in England, except that change came even slower to the colonies than it did to the mother-land. For example, orchards planted with seed-grown rather than grafted trees were the rule in America until early in the nineteenth century. Yet oddly, the apple was more important in the colonies, especially in the north, than it had ever been in Britain.

Planting apple orchards was among the first tasks the early settlers undertook. The first orchard in Massachusetts was planted around 1625 by a clergyman named William Blaxton (or Blackstone), who owned a farm on Boston's Beacon Hill. He also had the distinction of planting Rhode Island's first orchard as well. He had moved to Pawtucket, in 1635 and immediately set about planting another orchard. These early orchards were planted with imported seeds, and few if any attempts were made to domesticate America's native crab apples until some two centuries later.

Blaxton apparently was a cut above the typical American orchardist of later days in that he developed and propagated a type of sweet yellow apple that for years was known as Blaxton's Yellow Sweeting.

William Endicott, the first governor of the Massachusetts Bay Colony, also was an orchardist of some distinction. In 1648 Endicott traded 500 three-year-old apple trees for some 250 acres of land. His account book also contains an entry to the effect that his children had set fire to land near his apple nursery and destroyed another 500 trees. Clearly, Endicott dealt in apples on a considerable scale.

THE FUNCTIONS OF THE FRUIT

The reasons for the apple's importance in America were many, but very high on the list was cider—not the sweet cider we guzzle each autumn, but the fermented beverage we call hard cider. For despite the fact that America had in its infancy some of the best drinking water on earth, the colonists were quite leary of drinking it. As recent arrivals from Europe, they had reason of sorts for not drinking water. Sanitation in Europe was abysmal during the days of the colonies. In Elizabethan England, for example, it was standard procedure when someone died of plague for his survivors to throw his bedding into the nearest lake or river. For the most part, water in Europe was so contaminated that it wasn't worth one's life to drink it. Of course, people didn't understand why this was so, because the discovery of bacteria and the bacterial theory of diseases were many years away. All they knew was that disease was somehow connected with the water they drank, so they drank as little of it as possible. When Europeans came to America, they had no particular reason to forsake that belief.

The only alternatives the colonists had were to brew and drink beer, and to make and drink wines of some sort or another. (Cider, as a fermented fruit juice, is technically a type of wine.) Grape wine could be made from native grapes, but the flavor of this wine found little favor with the colonists. Attempts to introduce the European wine grape into the colonies failed dismally. There was a type of plant louse living in eastern North American soils that destroyed the roots of the European vines. The native varieties had developed sufficient resistance to this pest to survive despite it, but the early colonists did not understand the nature of the problem. Had they grafted the European vines onto North American roots, they would have met with greater success. But they didn't, and cider became one of the most popular drinks in colonial America.

Since the first American orchards were planted chiefly for cider, there was no great pressure to improve the varieties grown, nor was any great amount of time devoted to pruning or any of the other sophisticated techniques of apple culture. However, new varieties of apples spring up spontaneously from seed, and the seedling orchards of colonial America produced some outstanding varieties, such as the Roxbury Russet, the Rhode Island Greening, and the Newtown Pippin, all in cultivation by the first quarter of the eighteenth century.

During that same century the techniques of grafting and propagating trees became more widespread among the colonists, and at least some trees of specific varieties were cultivated along with the seedling trees. This trend away from the so-called natural trees toward grafted trees thereafter gained momentum.

The type of tree notwithstanding, the apple played a major role in American life from the outset. The popularity of cider drinking has been mentioned already, but cider was itself the raw material for other products that were essential to the colonists.

For one thing, cider can be easily transformed into cider vinegar. The change will happen spontaneously if too much air gets into the cider barrel during fermentation. Otherwise, a barrel of cider mixed with a third of a barrel of water was fitted with a loose-fitting lid and inoculated with yeast. In three to four weeks, the yeast would convert the alcohol in the cider to acetic acid, the sour principle in vinegar. The cider was diluted to prevent the end product from becoming too strong.

The vinegar was used in pickling, one of the important methods of preserving vegetables and fruits for winter use before the invention of hermetic canning. Recipe books well into the nineteenth century devoted page after page to various kinds of vinegar pickles, and some of them still find favor today, as the recipe for Pickled Apples (page 224) will testify.

Apple cider also could be distilled to make a type of brandy popularly called applejack. When strong enough, the applejack could be used to preserve other fruits; peaches, plums, and cherries were the most likely candidates for this kind of treatment. Beyond that, strong spirits, including apple spirits, found plenty of use in early medicine as antiseptics, anesthetics for surgery, and sedatives or stimulants, depending on the size of the dose.

The old-fashioned way of making cider was a lot more complicated than the definition, "the fermented juice of apples," might imply. First, the apples were crushed with a huge stone wheel that reduced them to a fine pulp, called pomace. This pomace was then stored in a huge open vat for twelve to twenty-four hours, depending on the weather.

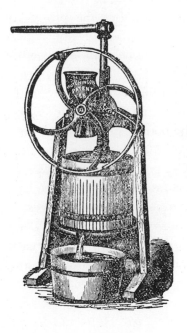

Hutchinson's Family Mill, as this nineteenth-century machine was called, was a combination fruit mill and press designed for making both cider and grape wine. The large wheel, turned by hand, powered the mill or grinder section of the device. The straight handle at the top was used to turn down the screw of the press section. From *The American Farmer's Hand-Book,* edited by F. W. O'Neill and H. L. Williams, New York, 1880.

The higher the temperature of the air, the sooner the apples would be ready for pressing.

Letting the pomace stand for a time sweetened and enriched it and gave the cider a good rich color. Exposure to the air worked these changes, and to further aerate the pomace the cider maker would turn it several times with a wooden shovel made especially for this operation. Then the pomace was formed into cakes called cheeses and hauled to the cider press.

The press could best be described as a little shed with posts instead of walls. Only the floor was slanted, so the apple juice would pour off one edge, and coming down from the roof was a large wooden screw that pressed against the apples when the cider maker turned it. The floor of the press was covered with straw, then a piece of "cheese" was placed on it. Then another layer of straw was added, then another piece of cheese, until the pile reached a height of about four feet, which might take a dozen layers of cheese. The top of the pile was covered with boards, and the screw pressed against these.

The juice was collected in a vat placed under the low side of the slanted floor. Afterward, the juice was poured into clean barrels, which had to be kept filled to the bung hole throughout the process of fermentation. This meant topping off the barrels each day to replace any liquid pushed out of the bung hole as the fermenting juice seethed and bubbled, for if too much air got at the cider, it would turn to vinegar.

Weeks later, when the cider stopped frothing (or "working"), the barrels were bunged up tight and allowed to rest. Slowly, any bits of pomace left in the juice would settle to the bottom of the barrel. Now the cider was ready for racking—drawing off the clear liquid into clean barrels and leaving the sediment behind. This process was repeated once more before the cider was left to age and mature, like a grape wine.

During the second racking, some farmers added another step, which was aimed at killing any bactria or yeast left in the cider. These organisms could cause the cider to sour. First, a few gallons of clear cider would be put into a clean barrel. Then the farmer would take a strip of rag that he had dipped in melted sulfur. He would set fire to the sulfur and lower the strip into the barrel, catching the tag end of it with the bung, which he pressed tightly into place. He would let the

In the scene above, two men grind apples into pomace, readying the fruit for the cider press below. The barrel (below, left) was the sort usually used for fermenting cider. From John Worlidge's *Vinetum Britannicum, or a Treatise of Cider*; London, 1676. (*Courtesy the Library of the New York Botanical Garden*)

sulfur burn until it had used up all the air in the barrel and had smothered itself. Then he would rock the barrel gently so the cider would absorb the fumes. He then removed the bung and the sulfur rag and filled the barrel with cider. The minute quantity of sulfur put into the cider this way would kill any harmful bacteria or yeasts, but it left no noticeable taste.

Cider making took a lot of time. The pressing was done in late October or early November. The first racking was done in late December, the second sometime in February, and the finished product wasn't ready for drinking much before June. However, if all went well, the cider would last until the next batch was ready.

Beyond cider and its by-products, the apple also found use as food for livestock, which were allowed to graze in early orchards, and, of course, as food for the colonists themselves. The expression "as American as apple pie" wasn't the product of an overzealous imagination. Apple dishes of one kind or another could be found at practically every colonial meal, especially in New England. The apple was made into pies and fritters and puddings and slumps, literally a host of dishes. The colonists had inherited some of their taste for apples from the British, along with many of the British recipes, but many other dishes were the products of American invention. A real taste of the period comes to us via the recipes for Squash and Apple Pudding (page 198) and Beef Mincemeat Pie (page 244).

PUTTING APPLES BY

The apple had one quality that made it especially popular in the early American kitchen. If handled gently and stored with care in October, a Rhode Island Greening might keep until the following March, a Roxbury Russet until June, and a Newtown Pippin until July. Certain apples would keep that long without the pickling, drying, or preserving required by every other fruit.

The colonial methods of storing apples were just about the same as those described by the Roman author Pliny. Apples were usually stored in cellars, sometimes buried in fine white sand, sometimes piled in heaps on the floor, then covered with straw, and sometimes packed in barrels with a little straw between each layer of fruit. On occasion, especially fancy apples would be hung from the cellar beams from pieces of string tied to their stems.

Apples that would not keep in storage (early or damaged fruit) could be dried quite readily by any one of several methods, such as quartering and drying in the sun or in a warm oven. A very common method, it seems, was to slice the fruit rather thin, then pass a piece of string through each of the slices, like so many beads. The strings would be hung among the rafters of the house to dry. In winter the dried apples could be soaked in water, then made into pies. One of the most popular dried-apple dishes was Schnitz und Knepp (see recipe on page 184).

Often, apples for drying were peeled, and the peels and cores were dried separately for later use in brewing a kind of beer. The beer was drunk, of course, but it provided something beyond good cheer. The froth that rose to the surface of the beer as it fermented was rich with yeast that could be used in making bread. Before the introduction of commercially prepared yeast cakes and dried yeasts, every household had to maintain its own yeast supply. The beer barrel was a good source.

Apples that were not sound enough for storage were often made into apple butter, or apple sauce, as it was frequently called. The process began with the boiling down of a quantity of sweet cider to about half of its original volume. The apples were peeled, cored, cut up into small pieces, and added to the concentrated cider. If quinces were available, a few of these would be thrown into the pot. A little spice might be added, then the mixture was put over a very low fire to simmer until it reached the desired consistency. The cooking could go on overnight, then the butter would be packed in earthenware crocks for winter use. An up dated version of this process can be found on pages 225–226.

EARLY AMERICAN CULTIVATION

By the end of the eighteenth century, seedling apple orchards were very nearly eclipsed by orchards planted with grafted trees. In Virginia, this process was completed somewhat earlier than it was in the northern states. George Washington had grafted espalier trees on his estate, Mount Vernon, and as early as 1686, a gentleman in Westmoreland County, a Colonel William Fitzhugh, reportedly had an orchard of 2,500 apple trees, most of them imported English varieties grafted onto seedling stocks.

Most of the apples of early Virginia seem to have been English imports. At the outset, growing apples from seed had been something of a necessity in the colonies. Apple seeds were small, light, and required no special care during the long sea voyage from Europe. As long as the seeds were kept dry, odds were they would be viable when planted. Transporting live trees or viable scions cut from trees was a far more difficult and costly proposition. The later resistance to improved methods of apple culture was probably a combination of a lack of real necessity, ignorance, and just general human reluctance to accept change.

Still, America's long romance with seedling orchards had one very positive result. As mentioned before, an apple tree grown from seed is a new variety of apple. Many, many thousands of seedling trees were tested in the apple-growing regions of America, and many hundreds of trees passed the test with honors. Because these trees were native to America, they were better suited to growing conditions here than most of the imported varieties. And since there were so many varieties to choose from, Americans were able to pick those that bore fruit as fine as any grown in the Old World.

After the Revolution, grafting and other sophisticated orchard techniques became common knowledge, and they were put to the task of spreading the country's finest varieties throughout the land. At the turn of the nineteenth century, out of the hundred or so varieties offered by professional nurseries the vast majority were native to America.

There was a factor other than national pride that fostered a sort of professionalism among American apple growers. In the earliest days, almost everyone lived on a farm or had enough land in town to grow his own fruit, but as the cities grew, they spawned new industries, and one of these was fruit growing. During the nineteenth century, the Hudson River, for example, was flanked with orchardists, who sent their apples by boat to the markets of New York. During the 1830s and

1840s, a barrel of Newtown Pippins might fetch as much as four dollars in New York, and that didn't include the price of the barrel.

Philadelphia was supplied by Burlington County, New Jersey, another area where apple growing was a fine art. Burlington was home to William Coxe, who was the foremost American apple grower of his day. He could honestly say, as he did in 1817 in a book he had written, that he had been "actively engaged in the rearing, planting and cultivating of fruit trees on a scale more extensive than has been attempted by any other individual in this country." Coxe was a devoted amateur fruit grower and a merchant by trade. His wealth allowed him to collect in his orchard practically every American apple tree available in his day. He and others like him made apple growing a science in this country.

During the nineteenth century, the population began to expand into more northerly districts than had been inhabited to any extent before. A systematic search began for apples that would thrive in such cold climates. Apple trees were imported from Scandinavia, northern Germany, and Russia, and they were tested and cross-bred with American varieties. By the middle of the century, there were more than 500 recognized varieties in cultivation in the United States.

The population expansion in America brought about a similar expansion in agriculture. As the nineteenth century wore on, more and more of the country's forests were felled and replaced by farms and orchards. The clearing of the land, of course, had begun two centuries before, but one of the consequences of felling the forests began to be felt during this century.

THE PEST PROBLEM

When reading old accounts of American agriculture, one finds descriptions of the fertility of the soil, of the prodigious crops produced, but one seldom finds references to insect damage. Frost, hail, and wind damage are mentioned, but not insects. The cultivated plants and fruits introduced into this country were foreign to the diets of the native insect species. As long as their native foods were available in the forests, these insects stayed pretty much in the wilds. But when the forests were replaced by farms, these amazingly adaptable creatures accustomed themselves to the new conditions. In addition, insect pests from Europe were introduced into America, concealed among the live plants, seeds, and scions for grafting that the colonists and later generations brought with them. It took years sometimes for the foreign insects to become established, but they did.

Few insects were mentioned in the earliest literature. Andrew Jackson Downing's *The Fruits and Fruit Trees of America* mentions six insects as apple pests and describes several more in a general introduction to insect problems. Some of the methods he suggests for fighting the pests should be of interest to organic gardeners and others who don't care for chemical insecticides.

For night-flying moths whose caterpillars feed on foliage, Downing suggests small bonfires, which would lure the moths, attracted by light, to a fiery death. The modern adaptation of this technique can be found in various traps using electric lightbulbs.

For aphids, or plant lice, as they are sometimes referred to, he recommends a strong tobacco "tea," applied with a paintbrush. Tobacco-based insecticides are still on the market today and can be applied this way, although a spray device is much simpler.

For tent caterpillars, he advises removing the webs with a long stick having a sort of coarse brush at the end. For the home fruit grower, this is still the best advice. But the operation has to be carried out early in the morning, while the caterpillars are still inside their "tents."

For canker worms, Downing recommends wrapping a four-inch-wide cloth bandage around the trunk of each tree and coating the bandage with a mixture of tar and oil. This would catch the worms as they tried to climb the trunk from the ground. (Pliny and Cato had suggested the same.)

For insects in general, he suggests setting out glass jars partially filled with a mixture of water, molasses, and vinegar. Bugs would be attracted to the jars, which could be hung among the branches, then fall into the sticky liquid and drown. Traps based on this principle were common fixtures in rose gardens in the 1940s, when the Japanese beetle seemed to be at some sort of population peak.

Downing was a professional nurseryman and a landscape architect of the first rank, and he leaves today's reader with the impression that the insect problem was nowhere near as serious as it is today. He describes six apple pests and about 600 varieties of apples. Downing's brilliant career makes it hard to believe that his cursory treatment of the insect problem was a matter of ignorance. Further, his book was aimed at professionals and serious amateurs, so there is little reason to think he was glossing over an unpleasant topic.

When U. P. Hedrick penned his *Fruits for the Home Garden* in 1944, he remarked that more than twenty insects troubled the apple grower in America. He then went on to provide methods to combat the five worst pests, noting that the insecticide sprays he recommended would control the others at the same time. Dr. Hedrick was director of the New York Agricultural Experiment Station at Geneva, New York, and one of America's top experts on fruit and fruit growing. One could therefore assume his information was hardly distorted.

Now, some thirty years later, publications issued by the United States Department of Agriculture list more than thirty insects that prey on the apple. This doesn't add up to proof that the insect problem has worsened over the years, but it does certainly provide food for thought.

It appears that by clearing the land, which forced native insects to change their feeding habits, and by the accidental introduction of foreign pests, American farmers engendered a problem that seems to have worsened during the course of the nineteenth and twentieth centuries.

Of course, there are certain things about modern apples, or for that matter all modern fruits, that may make them more desirable to pests than wild plants. First off, there is the matter of size. Modern domesticated apples are huge, compared to the wild apples of America today. A wild apple weighs a half ounce or less. A smallish domesticated apple weighs four ounces, or eight times as much. A really large apple can weigh two pounds—sixty-four times as much as a wild apple. The bigger fruit attract more pests.

Secondly, the acidic tannin that laces wild apples, making them astringent and mouth-puckering when eaten, has been replaced or greatly tempered by sweetness in the cultivated forms of the fruit. The additional sweetness of the modern apple may have made it a more desirable food to a greater number of pests than its wild forebears had been.

Finally, apples have been grown in huge orchards, for the most part, since the earliest colonial days in America. Large orchards, through sheer numbers of trees and tonnage of fruit, add up to big hunks of bug bait. Once America's insects learned to include the cultivated apple in their diets, they became a problem and have remained one ever since.

Spray equipment was developed during the nineteenth century to combat insect pests and plant diseases, and the use of chemical insecticides and fungicides became more common as the century drew to a close. The systematic search for better apple varieties continued, and more modern production methods came into use. But at the start of the twentieth century, apple growing in America still had an essentially nineteenth-century flavor.

THE TWENTIETH CENTURY

There were several forces at work that finally brought apple culture to its present form in America. The growth of the American cities fostered the creation of the professional fruit grower. Now the growth of the suburbs began forcing the growers further and further from their markets. The smaller growers were slowly forced out of business.

Cold storage was added to the picture, along with modern trucking. This tended to narrow the number of apple varieties in cultivation. Before, orchards had been planted with varieties of apples that ripened successively from July through the winter months. But many of these varieties were not the best producers in volume of fruit. Now with cold storage and quick, dependable transportation, the growers could concentrate their efforts on trees that produced heavily and reliably.

At the same time, and even a little earlier, something else was happening. Professional nurserymen had long supplied orchardists with trees, but it had always been something of a local industry, with local nurseries providing locally proven varieties to local orchardists. But as transportation improved, big regional nursery firms evolved, and quite naturally they pushed the varieties they had developed and whose propagation they controlled.

The preeminence of the large nurseries, which could ship dependable trees throughout vast regions, helped to further narrow the number of different apple varieties in cultivation. This process was further abetted by the canning industry and finally by the evolution of supermarket chains, whose buyers wanted only apples they would be able to sell to a mass market.

The 1872 edition of Downing's *Fruits and Fruit Trees of America* listed more than 1,000 varieties of apples. In 1892, L. H. Bailey compiled a list of apple varieties available from ninety-five nurseries whose catalogs he had collected. His tally was 878, but some of these were duplications and crab apples. H. P. Gould of the USDA checked Bailey's list and came up with the figure of 735 apple varieties and 40 crabs. Then Gould did some compiling of his own, based on 100 catalogs issued in 1910. Gould discovered that in only eighteen years, the number of apple varieties available had diminished by nearly a third, from 735 to 472. Gould did another such survey in 1941, and the number now had fallen to 269. Today, the number hovers around 100, about 10 percent of the number of varieties in cultivation a little more than a century before.

Of course, there is no question that many of the varieties that have slipped into oblivion deserved nothing better. However, many of the now-forgotten apples were of excellent quality. The market today is dominated by sweet varieties such as Delicious, Yellow Delicious, and McIntosh, and if your taste preference is for tart fruit, it often is difficult to find a satisfactory apple.

With today's production costs apple growers cannot afford to concentrate on anything but sure sellers and on methods that will produce acceptable quality fruit that can be sold at competitive prices. Most of the efforts of the growers and plant scientists now seem aimed at improving the few varieties still in cultivation. Developing redder red apples has been one of the principal goals of the developers of the Red Delicious apple for the past half century. Very red apples find greater acceptance among consumers so far removed from the farm that they don't realize that an apple with a green skin can be fully ripe and delicious.

Other efforts have been devoted to the development of safer and more effective pesticides, fungicides, and other disease-control agents. Work has also centered on the development of chemical plant regulators, which hasten fruit budding and ripening, and on chemical thinning agents, which help the trees to cull themselves of excess fruit in years when they produce more than they can carry to full growth. In the past, this task had to be done by hand.

Finally, with land prices on the rise, researchers are concentrating on ways of getting greater fruit yield per acre through effective use of fertilizers and other plant nutrients and on the development of fruit trees that can produce more fruit in less space. Until recently, most of the work in this latter area has centered on grafted dwarf trees, with scions from standard-size trees grafted onto rootstocks that inhibited growth.

In the 1950s, research took a turn with the discovery of what are called spur-type sports. These were the products of minor mixups in the genetic material of single fruiting branches. This type of change doubtlessly had been going on as long as there had been apples, but apparently nobody noticed or cared. But the result of this mutation was that from the point of the change on, the branch produced its fruiting spurs (the places where the fruit grows) much closer together. In other words, the mutant branch would produce the same amount of fruit as the normal branch but would do so using only half its length.

What was even more important was the fact that when such a branch was grafted to a rootstock, the result was an entire tree with the same characteristics. The spur-type dwarf is considerably smaller than the standard tree, yet it produces a comparable quantity of fruit. An important side benefit is that since the upper portion of the tree is a dwarf by nature, there is no need to graft to a dwarfing rootstock. Many of the dwarfing rootstocks have the disadvantage of being more prone to injury than standard rootstock.

Spur-type dwarfs have been found among many of the major varieties in cultivation today, and if they are combined with dwarfing rootstocks, the apple orchards of tomorrow may well look like fields full of hedgerows, except that the "hedges" would be covered with apples.

2

The Apple and Its Varieties

A cluster of apples from Hendrik Cause's *De Koninglycke Hovenier,*
Amsterdam, 1676. (*Courtesy the Library of the New York Botanical Garden*)

The Apple tree hath a bodie or trunke commonly of a meane big-
nesse, not very high, having long armes or branches, and the same dis-
ordered: the bark somewhat plaine, and not very rugged: the leaves be
broad, more long than round, and finely nicked in the edges. The flowers
are whitish tending to a blush color. The fruite or Apples do differ in
greatnes, forme, colour and taste; some covered with a red skin, others
yellow or greene, varying infinitely according to the soyle and climate;
some very great, some little, and many of the middle sort; some are

[33]

sweate of taste, or something sower; most be of a middle taste between the sweete and sower, the which to distinguish I thinke it inpossible; notwithstanding I heare of one that intendeth to write a peculiar volume of Apples and the use of them; yet when he hath done that he can do, he hath done nothing touching their severall kindes to distinguish them. This that hath beene said, shall suffice for our historie.

Thus Jean Gerarde, in his book *The Herball*, described the apple tree and its fruit nearly 400 years ago. But Gerarde notwithstanding, we intend to present a discussion of apple varieties, distinguishing their several kinds and providing a more detailed description of the tree and its fruit.

The Apple Seed and How It Grows

As a plant, the apple tree is classified as a member of the rose family. The small, applelike hips found on a rosebush in winter point to the relationship, and a comparison between a wild rose and an apple blossom should clinch the argument.

Actually, the rose family includes a very large share of fruits grown in this country: the apple and crab apple, the quince, pear, cherry, plum, peach, nectarine, apricot, blackberry, raspberry, strawberry, dewberry, and even the almond. So Pliny wasn't so far from the truth after all when he called peaches "apples."

The relationship between the rose and the apple might be further understood as one notes some of the European "roses" that have found their way into the kitchen, such as the medlar and worden, the sloe or blackthorn, the rowan or mountain ash, the service pear or sorbus, and the hawthorne, beam, and cotoneaster. But that is just the beginning, because the rose family includes about 2,500 members worldwide, and many produce edible fruit.

One could even mention the rose itself. Its petals have been eaten in the form of fritters (the same applies to apple blossoms), cooked with capons, and added to wine. Rosewater was a popular flavoring extract for centuries, and rose hip tea still has its enthusiasts.

The apple tree itself, as Gerarde remarked, is of average size as trees go, seldom reaching a height of more than 40 feet if left to grow freely. By way of contrast, many an oak tree has grown to the height of 100 feet, and a few have gone as high as 150 to 180 feet. The bark of the apple tree tends to be on the smooth side, but it can be quite rough, as in the case of the Newtown Pippin.

When left unpruned, the tree will form a round head, which causes problems for the fruit grower. When the head is left round, the lowest branches are in shade and won't produce fruit because of it. Standard orchard practice for years has been to cut back or prune the branches, so the tree will have the shape of a rough pyramid or cone, with the point at the top, of course. That way all the branches get the same amount of sunlight.

As Gerarde noted, the branches don't spring from the trunk in any particular pattern. They simply fill in the available spaces to form a round head. Pliny said the apple's branches grow like whiskers on the muzzles of certain animals, and that is a pretty fair description of some apple trees.

Each spring, a good-sized tree produces from 50,000 to 100,000 apple blossoms, most of them grouped in clusters of five or six. Most of the clusters spring from the ends of short, woody stems called fruit-ing spurs. Occasionally single blossoms will appear along the length of the branch, especially where the wood is young and the bark still quite thin.

Each blossom has five petals, usually white but often with a pinkish tinge. The petals are the most conspicuous parts of the blossom, but if one takes a closer look, he will find there is a lot more to an apple blossom than petals.

Just imagine for a moment that we were going to build a giant model of an apple blossom. We might start out with a wine bottle, one of those straw-covered Italian chianti bottles. We would peel off the straw and we would be left with a sort of glass teardrop with a long neck. We would glue the base of the bottle to a broomstick, and this would be the stem of our model flower.

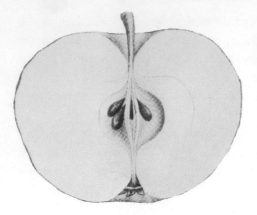

The Red Astrachan, shown here in cross section, was one of several varieties introduced into America from Russia during the nineteenth century, as growers searched for apple trees capable of withstanding severe winters. A summer apple, ripening from July to mid-August, it is found only in limited cultivation today. From a lithograph published by R. H. Pease, Albany, New York, circa 1850.

Next, we would glue five huge petals around the base of the neck, at the point where the bottle begins to grow fat. And we would curve the petals upward, so they formed a sort of shallow dish around the neck of the bottle. Beneath each petal we would attach a smaller, triangular piece with its point directed outward. These would be the sepals that covered the blossom before it opened.

To finish the job, we would fasten two or three pipe cleaners at the point where each petal joined the bottle. We would let the pipe cleaners stand up tall but would trim them off so they would be the same height as the neck of the bottle. At the end of each pipe cleaner, we would put a little knob. Finally, we would divide what we might call the body of the bottle, the fat part below the petals, into five vertical compartments, and inside each compartment, we would put two tiny peas.

In the language of botanists, the body of the bottle would be called an ovary. Each compartment inside would be a carpel, while the peas within each compartment would be ovules. The neck of the bottle would be called the style, and the mouth would be the stigma. The entire bottle would be the pistil. These are the female parts of a flower.

Many flowers are both male and female, and the apple blossom is one of them. The male parts in our model are represented by the pipe cleaners, which would be the stamens, while the knobs would be anthers. And inside each anther would be tiny grains of powder, the pollen.

The pollen grains are analogous to the "seed" of male animals, while the ovules correspond to the eggs of females. As in animals, the male cells must fertilize the female cells, or else there will be no possibility of offspring. Of course, the analogy to animals only holds so far, because plants "give birth" to seeds, which in turn grow into "offspring."

The actual process of fertilization in the apple goes something like this: A grain of pollen adheres to the stigma at the very top of the pistil. Soon the pollen grain begins to act like a tiny seed. It germinates and sends a "root" through the tissue of the stigma, down through the entire length of the style and right into one of the ovules.

In the middle of the ovule is a tiny body, which might be thought of as a half cell, because it contains only half the number of chromosomes, half the genetic material of a whole apple cell. This half cell is called the egg nucleus. When the "root" of the pollen grain reaches the ovule, it releases its own genetic material, which then combines with that of the egg nucleus. In this way, the "egg" is fertilized in the apple blossom, and the "fertilized egg" grows into an apple seed.

The description has been simplified, but the information omitted is not germane to our purposes. What is important here is the fact that the apple gets half of its genetic material from its "father" and half from its "mother." The explanation should be clear: The "father" is the plant that produced the pollen, while the "mother" is the plant that produced the egg nucleus.

However, we also said that the apple blossom has both male and female parts. Could it be that the same apple blossom is both "parents" to its seeds? The answer is that this is possible in some varieties of apples, but for the majority of apple varieties, the answer is no. Most apple trees have a chemical mechanism for rejecting their own pollen and pollen produced by other trees of the same variety. In most cases, there must be cross-pollination between two or more varieties.

Of course, the question arises, how does the pollen of one apple tree get to the blossoms of another. Pollen is as fine as dust, and wind can carry it from tree to tree. But the most reliable way is via honeybees. The bee's body is covered with tiny hairs, and as she goes into the blossom to get the nectar produced at the bases of the petals, her body brushes against the anthers, and the pollen sticks to the hairs. At the same time she also brushes against the pistil and leaves some pollen behind. So as she goes from tree to tree, she takes the pollen of one to the pistil of the next. Without the honeybee, the apple crop would fall to nearly nothing, and the same would happen to almost every variety of fruit and vegetable.

Since the pollen comes from one tree and the "egg" from another, apple trees don't "grow true" to variety from seed. But there is an underlying reason that makes this practically impossible. The apple is a hybrid, which means it draws its genetic material from many different species of wild apples. That genetic material is divided up in the pollen grain and in the ovule to form the two half cells that become the apple seed. There is no easy way to explain precisely what happens, but let it suffice to say that there are thousands of bits of genetic material in each cell and they can be divided in an almost countless number of ways. In all probability, no two apple seeds have ever had the same genetic makeup, and that applies to the tree that grows out of each seed and to the fruit each tree produces.

Now if each pollen grain fertilizes one egg nucleus and there are ten ova, each with an egg nucleus, then it follows that ten pollen grains are necessary to completely fertilize the blossom. If this doesn't happen, if the blossom is poorly fertilized, the entire blossom falls from the tree. If all or most of the "eggs" are fertilized, the petals drop and what remains develops into an apple.

When we constructed our mental model of an apple blossom, we said that the chianti bottle represented the female part of the flower and that the fat body of the bottle was its ovary. Now imagine that glass could grow, because this is essentially what happens to the ovary after the "eggs" are fertilized. The walls of the ovary grow thicker and thicker and form the flesh of the apple. By the time an apple reaches its full size, the ovary walls weigh hundreds of times more than they did at the outset.

All that remains of the sepals, the stamens, and the style of the pistil is that little bunch of dried material in the dimple at the bottom of the fruit. The petals, which dropped off long before, had acted principally as bee bait.

Nine apples drawn from life by Johann Hermann Knoop. From Knoop's *Pomologie*, Amsterdam, 1771. (*Courtesy the Library of the New York Botanical Garden*)

No apple tree could bear as many apples as it does blossoms. If that happened, the average tree would produce hundreds of bushels of fruit. The actual yield is somewhere between fifteen and thirty bushels, depending on the size and vigor of the tree. As mentioned before, poorly fertilized blossoms fall from the tree. Later on, the tree may even drop fruit that has already formed if the crop is too heavy, or if too many blossoms were fertilized. In the average year, no more than 2 to 5 percent of the blossoms develop into apples. But the apple tree's system of fertilization is so inefficient in a sense that if the tree didn't produce that many flowers it would never bear a decent-sized crop.

There is no need to describe how the apple grows beyond the point we have already reached. The apple gets bigger, and its final size, shape, and color, as well as its flavor and aroma, are determined by the genetic structure of the "mother" tree. The seeds inside the apple could have been "fathered" by one or ten different trees, but it makes no difference. The ovary walls that form the apple's flesh are part of the "mother" tree, and "her" genes alone determine the qualities of the fruit. However, the seeds have their own mixtures of genetic materials, so when they grow into trees, they will produce fruit according to their own genetic makeups.

This, finally, is why when each apple seed is planted, it produces a new variety of tree and fruit. And this is why grafting or some other artifical means of propagation is necessary to produce many apple trees of the same variety, if that is desired.

When the apple has reached its full size, the ripening process begins. All during its growth period, each apple has drawn nutrients from the "mother" tree. The stem of the apple has acted as a tiny pipeline, carrying food to the growing cells. But a few weeks before harvest time, and this varies greatly according to the variety, a group of cells grows across the base of the stem, at the point where the stem connects to the branch. These cells block the "pipeline," cutting off the apple from any further nutrition from the tree. Although the apple isn't ripe yet, it has already been isolated from the tree.

Then a remarkable thing starts to happen. The apple, even though isolated from the tree, continues to go on living, and to do this, it begins to use up energy stored in its own flesh. All during the growing season, when the pipeline to the tree was open, the apple's flesh was filled with starch. Now the apple starts to convert that starch into sugar, gaining a certain amount of energy from the conversion. The apple also starts to produce red and yellow pigments, which will give the fruit its final color. The pigments are the same as those that turn leaves red and yellow in autumn. The seeds inside the apple, which have been white all along, turn a rich red brown. The apple is then ready for picking.

Even after the apple is picked, it goes on living. Once all the starch has been converted into sugar, the cells begin to consume the sugar itself. Eventually, the sugars run out, and the cells begin to die. Decay sets in.

Knowing the way an apple ripens makes it easy to see how a mature fruit can be picked and allowed to ripen off the tree. Once the apple has reached full color, which is dependent upon sunlight, it can be picked and put anywhere to ripen. The pipeline to the tree has long since been shut. More than that, if the apple is stored in a cool place, the process of ripening and decay can be slowed. This is why a Newtown Pippin can go into the cellar in October, reach its peak of ripeness in March, and still be quite palatable in July.

Now on to just a few of the thousands of apple varieties that have been cultivated in America. Some are cultivated widely by commercial growers; others still find their way into home orchards. A few have been included because of their importance in history, whether planted today or not.

Baldwin

The Baldwin apple originated as a chance seedling in Wilmington, Massachusetts, around 1740. The tree was discovered in 1793 by Samuel Thompson, who called it to the attention of Lieutenant Colonel Loammi Baldwin of the 38th Infantry Regiment of the Continental Army (Ret.). The colonel propagated the tree and introduced it throughout Massachusetts. At first, it was called the Pecker or Woodpecker apple in honor of the birds that visited the tree, which was at least a half century old when discovered. It was given its present name years later to

This old European apple, the Carmelite Reinette, is shown here in a German bookplate dated 1802. The term *reinette* is derived from the Latin *renatus*, meaning "reborn," and was applied to apples propagated by grafting. The term *pippin*, in contrast, was applied to an apple grown from a pip or seed. From Johann Volkmar Sickler's *Der teutsche Obstgärtner*, Weimar, 1794–1804.

honor the memory of the colonel, who was with Washington on the Christmas night in 1776 when the Americans crossed the Delaware to attack the Hessian mercenaries quartered at Trenton.

The Baldwin apple was the most popular variety in the Northeast through most of the nineteenth century. As late as 1915 it was the most widely produced variety in the country. In that year, Baldwins accounted for more than 13 percent of the total American apple crop.

The fruits of the Baldwin are large and roundish, narrowing just a little toward the blossom end. The skin is yellow but striped and nearly covered with crimson on the side that faces the sun. Baldwins are good keepers, ripening in the cellar between November and March. The tree is vigorous, highly productive, and long-lived.

Baldwin apples still find their way to the market and are excellent for dessert or for cooking. Baldwin trees can be purchased from most nurseries.

Ben Davis

Little is known about the early history of this variety. It probably originated around the turn of the nineteenth century in Todd County, Kentucky. Whatever its origins, Ben Davis dominated the apple-growing regions of the South the same way that Baldwin dominated the Northeast.

The fruit is large, roundish, narrowing toward the blossom end. The skin is marbled with bright red on a yellow ground. The flavor is mild, and the apple is not particularly noted for its quality. However, the fruits are extremely firm and keep well with minimal care. The tree is vigorous and very productive but is susceptible to nail-head canker, a disease that usually kills it. The fruits occasionally are found in markets in the South.

Cortland

The Cortland was introduced in 1915 by the New York Agricultural Experiment Station at Geneva. The original tree grew from a seed produced in 1898 by crossing Ben Davis with McIntosh. The Ben Davis was the "mother" tree. The fruits ripen later than McIntosh, which they resemble in size and shape. The skin is yellow with a red cheek. The flavor is sweet and suitable for dessert or for cooking. The Cortland is especially useful for salads (see recipes for Spanish Salad, page 164, and Ham and Apple Salad, page 165), because its flesh is slow to turn brown after it is cut. When stored under cellar conditions, Cortlands are at their best from October into December.

Delicious

About one in four apples grown in America today bears the name Delicious, and it is produced by trees descended from a chance seedling found in 1872 on the farm of Jesse Hiatt of Peru, Iowa. That the original tree survived at all was something of a miracle.

Hiatt had an orchard planted with trees of a known variety. Since the seedling was a tree of unknown parentage, growing wild, he twice cut it down. Each time it grew back. Hiatt finally decided to tolerate the tree and see just what kind of fruit it might produce. Since the seedling had sprung up near a Yellow Bellflower, Hiatt guessed that the latter might have been its "maternal" parent.

The Yellow Bellflower had originated in Burlington, New Jersey, sometime around 1800. William Coxe was the first to describe it. The fruit had a pale yellow skin and was large and oblong, tapering toward the blossom end. In 1882, Hiatt's "weed" finally bore a single fruit, and the fruit bore out Hiatt's suspicion: The apple was shaped almost exactly like Yellow Bellflower, but as it ripened, the seedling's fruit turned red. Hiatt tasted the apple, liked it, and decided to promote it as a new variety, which he dubbed Hawkeye.

He entered the fruit in several Iowa fairs, but it failed to gain any recognition. Then in 1893, he sent samples to a competition sponsored by Stark Brothers Nurseries & Orchards Company. The fruit attracted the attention of C. M. Stark, the president of the firm, but the papers identifying the apple's owner had gotten lost. A persistent man, Hiatt reentered Hawkeye in the 1894 Stark Brothers competition, and this time contact was made.

Stark Brothers bought the rights to propagate and distribute Hiatt's Hawkeye and renamed it Delicious. The firm was convinced it had a winner and spent $750,000 on introducing and promoting Delicious. From an insignificant share of the total United States apple crop in 1900, Delicious had risen to an unquestioned first place in 1942 (the first year of USDA records), when it accounted for nearly 17 percent of the total American crop. Delicious has remained the number one apple in the country ever since.

The fruits of Delicious are large, red, and oblong, tapering toward the blossom end. The base of the fruit is marked by five conspicuous knobs. The flesh is crisp, fine grained, and juicy; the flavor is sweet and mild. This apple is eaten fresh in salads or out of hand but finds no use in cooking. The fruits ripen in October and keep reasonably well. They have a tendency to become mealy if at all overripe. The trees are moderately vigorous, produce heavily, and are fairly resistant to winter cold. Delicious trees, especially the newer "super-red" types, are available from most nurseries.

Early McIntosh

This fruit also was introduced by the Agricultural Experiment Station at Geneva, New York, in 1923. It was produced by crossing Yellow Transparent with McIntosh. Once again the McIntosh was the "father," or pollen producing tree. The fruit, which ripens in early August, is similar to the McIntosh but smaller. The skin is red all over, and the flesh is juicy and sweet. This apple can be used for dessert or for cooking. The trees are vigorous, hardy, and reliable annual producers. Trees are available from many nurseries.

Golden Delicious

Golden Delicious appeared as a chance seedling on the farm of Anderson H. Mullins of Clay County, West Virginia. The actual parentage of the tree is unknown, but research indicates that the original seed was produced by a Grimes Golden pollinated by a Golden Reinette. Both varieties were present on the Mullins farm around the turn of the century, when the seed doubtlessly was spawned. (It takes about fifteen years for a seed-grown tree to bear fruit.) Mullins sold the tree for $5,000 in 1914 to Stark Brothers Nurseries & Orchards Company, which erected a steel cage around the tree (a small piece of Mullins's land went with it) and set about the business of propagating and introducing it.

Today, Golden Delicious is one of the most widely grown apples in America. The fruit is relatively long and tapered, like the Delicious apple, and the skin is a bright golden yellow, sometimes with a rosy cheek. The flesh is firm, crisp, with a delicate, almost pearlike flavor. Many people consider Golden Delicious one of America's finest dessert or eating apples. It is also good for cooking, but because of its sweetness the sugar specified in recipes often must be reduced a bit.

Golden Delicious ripens late in September and is a good keeper. The trees bear young and vigorously and are available from most nurseries.

Grimes Golden

Grimes' Golden Pippin, as it once was called, originated sometime before 1800. It was a chance seedling discovered on the farm of Thomas Grimes in what was then Brooks County, Virginia (now located in West Virginia). The fruit was highly prized as a dessert apple in the South and in the states that border the Ohio River.

The fruit is medium in size, shaped something like Golden Delicious, and is covered with a rich golden skin. The flesh is yellow, juicy, and crisp, with a tart, spicy flavor that earned Grimes Golden its popularity. The fruits are good for dessert and for cooking. Unfortunately, they are not the best keepers and should be used as soon as possible after they ripen in September. Grimes Golden trees are available, usually as dwarfs, from nurseries carrying old-time varieties.

Gravenstein

Gravenstein is a variety of European origin. Some say it is from the Jylland region of Denmark, while others name Schleswig–Holstein, the German province that borders it. Wherever this apple came from, it reached the United States very early. It was in California by 1820, when it was planted at a Russian settlement near what today is Bodega in Sonoma County.

Gravenstein is a handsome apple and a big one. The fruit is roundish but tending to be a bit lopsided. The skin is yellow and marked with bright red and orange. The flesh is tender, crisp, and highly flavored when the apple is grown well. It is a good eating apple and one of the best for applesauce and pies. The trees are moderately resistant to winter cold, but they have one characteristic that makes them unsuitable for modern commercial orchards: The apples are not ready for consumption all at once; they take a period of several weeks to ripen and drop from the trees. For the home orchardist this same characteristic can be a blessing, because it eliminates the problem of having to can, store, or eat the tree's entire crop at once.

The Gravenstein apple begins to ripen in September and will keep into November. Trees, especially dwarfs, are available from only a few nurseries.

Jonathan

The original Jonathan tree was found on the farm of Philip Rick of Kingston, New York. The first description of the apple was in an article presented in 1826 to the New York Horticultural Society. The

Lithograph published by R. H. Pease, Albany, New York, circa 1850.

author of the article was Judge J. Buell of Albany, who had introduced a variety of plum called Jefferson about a year before. Buell named the fruit Jonathan after Jonathan Hasbrouck, who first brought the fruit to his attention. But the Jonathan also was called the Philip Rick apple or the King Philip for many years after.

The Jonathan apple is medium sized, with a skin that is basically yellow but nearly covered with bright red stripes. The shape is round but tapered a little at the blossom end. The flesh is firm and white yet tender and juicy. The flavor is mild but sprightly. Overall, it is an excellent fruit for dessert and for cooking. Unhappily, this apple, which ripens in October, does not keep well.

The Jonathan tree is moderate in size and vigor. The trees bear heavily each year and are capable of partial self-pollination. The trees are highly susceptible to a disease called fire blight but fairly resistant to a more common disorder called apple scab. Jonathan trees are available from most nurseries.

Lodi

This variety was introduced by the New York Agricultural Experiment Station in 1924. The original seed was produced by crossing Montgomery with Yellow Transparent. The variety is quite hardy and disease resistant. The fruits are yellow and quite large, and as early apples, rather well flavored. (Generally speaking for most varieties, late apples tend to be more flavorful.) The fruits ripen by mid-August, and the trees are vigorous and heavy producers. Some fruit thinning may be necessary in years when crops are heavy. Lodi trees are available from most nurseries.

Macoun

Macoun was introduced by the New York Agricultural Experiment Station in 1923, the same year as Early McIntosh. The original seed was produced by pollinating McIntosh with Jersey Black. The fruit is large, all red, and highly flavored. Many people consider Macoun the finest dessert apple grown in the Northeast. It can also be used for cooking. Overall, Macoun can be regarded as a better quality McIntosh, which it closely resembles, but Macoun ripens later and keeps longer. Macoun trees are available from many nurseries.

McIntosh

In 1811, John McIntosh, who had a farm in Dundas County, Ontario, found a young apple tree growing amid some brush on his property. He transplanted the tree to his garden, and soon it began to bear the fruit that would someday make his name famous.

At first, McIntosh tried to capitalize on his luck by planting the seeds from the tree and selling the seedlings to neighbors. Then in 1835, he learned about grafting from a visitor to the McIntosh farm. After that, he grew seedling trees but grafted them with scions from his choice tree. At first the variety was known as McIntosh Red, but later the name was shortened.

Dundas County is not far from Niagara Falls, and it wasn't too long before the McIntosh apple spread into New York State and eastward into Vermont. By 1900 it had become fairly well established in the United States, and by 1915 McIntosh accounted for nearly 1 percent of the total U.S. crop. In terms of competing varieties, however, it was in twenty-eighth place. Today, McIntosh is in second place, behind Delicious, which has the lion's share of the market.

The skin of McIntosh is green, heavily striped, and marked with red. The flesh is tender, sweet, and highly aromatic. The fruits are medium sized and ripen in September. They are good as eating apples and for applesauce. The tree is very large for an apple tree, with spreading branches. It is hardy but somewhat susceptible to apple scab. McIntosh trees are available from most nurseries.

Newtown Pippin

The Newtown Pippin is one of the most celebrated varieties in American apple history. Its story began sometime after 1700 on New York's Long Island.

The Gershon Moore estate was in Newtown (known today as Flushing, Queens) within the city limits of New York. On the estate was a little swamp, and near the swamp stood an apple tree. One day, an apple fell from that tree, and it contained the seed that would become the Newtown Pippin.

The tree itself was a feeble grower, with unusually rough bark

The Black Gilliflower, or Sheepnose, was considered the best apple for baking during the eighteenth and nineteenth centuries. It first appeared in New England during the 1700s and was noted for its deep red, almost black, coloring. Although very popular in years past, it is rarely found today. From a lithograph by R. H. Pease, Albany, New York, circa 1850.

for an apple tree. But it bore heavy crops of green-skinned apples that people of the day thought were the finest ever grown. The apples were medium sized, roundish, but a little misshapen, because of two or three ill-formed ribs that ran from stem to blossom end. The fruit had a superlative flavor, ripening late in winter and keeping in tiptop shape right through June if stored in a cool cellar.

As early as 1759, the Newtown Pippin had reached England as the first American apple sent back to the motherland, a country noted for its own apples. Benjamin Franklin was there at the time, pleading the American cause before George III's privy council. Franklin had the apples sent to him in London, and he distributed samples to friends, even sent some to court. The Newtown Pippin found such favor in England that an export business was based on it, a business that survived for more than a century. A barrelful of Newtown Pippins sold for nine dollars in London during the 1840s.

But the Newtown Pippin had its flaws. The tree was finicky about soil conditions and would not tolerate a cold climate. Overall, it seemed to grow best in the Hudson River Valley. Yet despite its limitations, every amateur fruit grower wanted a Newtown Pippin tree in his orchard, and the original tree succumbed to excessive cutting in 1805.

Strangely for an apple of such prominence, the Newtown Pippin's history became confused in the nineteenth century. In 1845, A. J. Downing applied the name to what he regarded as two distinct varieties. He described a green-skinned "Newtown Pippin" and a "Newtown Pippin, Yellow," which had yellow skin and a red cheek. He felt that the two varieties were similar but that the yellow had a markedly smoother skin, a flatter profile, and a higher perfume, while the green was more juicy, crisp, and tender. He called both highly flavored and said they had similar keeping qualities.

Today, there is an apple called Yellow Newtown, which is cultivated to some extent in Hood Valley, Oregon, where the variety was introduced around 1900. This apple is said to be the same fruit as the Albemarle Pippin, which Thomas Jefferson planted on his Virginia estate, Monticello, in 1773. Some writers say the yellow-skinned Albemarle Pippin was a distinct variety, but others identify it as the green-skinned Newtown Pippin of New York.

The whole business became quite confusing, but U. P. Hedrick found the answer. Dr. Hedrick wrote in *A History of Horticulture in America to 1860* that he once grew all three "varieties"—Newtown Pippin, Albemarle Pippin, and Yellow Newtown—side by side at the New York Agricultural Experiment Station. When grown under identical conditions, the three proved identical. Apparently, soil and climate could affect the Newtown Pippin to such a degree that they could change the color and even the shape of the fruit.

Whatever the explanation, the Newtown Pippin was a fine old apple, though sadly, it is very difficult to find it today.

Northern Spy

The Northen Spy has one of the most tortuous lineages in the apple tribe, thanks to the family of humans involved in its development.

Oliver Chapin bought a farm in Ontario County, New York, in the year 1789. The following spring, Dr. Daniel Chapin joined his brother, bringing along some apple seeds, which he planted on Oliver's land. Then in 1796, a third brother, Heman, came to Ontario County, and in 1800, Heman bought a farm near Oliver's.

It is not clear whether Heman planted apple seeds of his own or took some of the seedling trees planted by Daniel Chapin some ten years before. However, Heman Chapin is credited with planting the first Northern Spy tree. What is peculiar is that the tree never bore fruit on Heman's farm. It died before it was old enough. However, the tree threw up some suckers from its roots before it died, and these were planted by Roswell Humphrey, Heman's brother-in-law. On the nearby Humphrey farm the Northern Spy finally produced the apples that would make it one of America's very important varieties.

As if to prove that lightning can strike more than once, the Northern Spy was one of many seedling trees that Heman Chapin planted in his orchard. That same planting of trees also produced Early Joe and Melon, two varieties highly regarded during the nineteenth century.

The fruit of Northern Spy is large, roundish, with basically yellow skin that is heavily striped with bright red. The flesh is fragrant, just a little tart, and blessed with a particularly fresh flavor. It is a fine apple for dessert or for cooking, and it keeps well, ripening in November and holding its quality well through February under cellar conditions.

While vigorous and hardy, the standard tree is slow to come into bearing. This problem is eliminated, however, in dwarf trees. The tree blooms later in spring than most varieties, which can lead to some pollination problems. However, if other late blossomers, such as Golden Delicious or Rome Beauty, are planted nearby, there is no problem. Northern Spy trees are available from a few nurseries.

Rhode Island Greening

U. P. Hedrick wrote in *A History of Horticulture in America to 1860* that this famous green apple was grown from seed in 1748 by a tavern owner named Green at a place called Green's End, near Newport,

Left: *"Malus Carbonaria longo fructu,* the baker's ditch apple tree." From John Gerard's *The Herball or Generall Historie of Plantes,* "very much enlarged and amended by Thomas Johnson," London, 1633.

Center: *"Malus,* the apple tree"; at right is *"Malus sylvestris,* the wilding or crabbe tree." From John Parkinson's *Theatrum Botanicum,* London, 1640. (*Courtesy the Library of the New York Botanical Garden*)

Rhode Island. There are enough *greens* in the story to jaundice the eye, but Hedrick is a reliable historian.

One thing seems certain, and that is the place of the Rhode Island Greening in American apple history. This fruit was one of the first American varieties propagated to any extent in colonial New England, and it was one of the first really good apples to come out of the New England states.

A. J. Downing remarked that this apple was so familiar that to describe it to his readers seemed almost superfluous. That was in 1845, and from 1900 until recently, with the introduction of the Australian Granny Smith apple, the Rhode Island Greening was one of the few green apples commonly found in the American market.

The fruit of the Greening is large, roundish, though a little squashed, and covered with a smooth, yellow green skin. (By way of contrast, Granny Smith is somewhat smaller and has a bright, grass green skin.) The flesh of the Greening is crisp, yellow, and finely grained; the flavor is rich, tart, and aromatic. It is an excellent eating apple, despite its unripe appearance, and is one of the finest of all cooking apples. The fruit ripens in late October and keeps easily until March.

The Rhode Island Greening tree is vigorous, moderately hardy, and very productive. Unhappily, it is a very poor pollen producer and must be grown in combination with two different pollen-producing trees—three different varieties in all. This way the two good producers can pollinate the Greening and each other. Rhode Island Greening trees are available from many nurseries.

Rome Beauty

The Rome Beauty, or simply the Rome apple, was a chance seedling that developed in a somewhat unlikely fashion.

Joel Gillett had a farm in Proctorville, Ohio, which was then part of Rome township. In 1816 he bought a number of apple trees from Putnams' Nursery in Marietta, which was some eighty miles to the northeast. This nursery had been founded in 1796 by Israel and Aaron Putnam, nephews of Major General Israel Putnam, who was second in command to Washington during the Revolutionary War. The Putnam nursery was the first west of the Allegheny Mountains.

The trees that Gillett had purchased were grafted trees, the first commercially available in that area, and when he went to plant them, he discovered that one had sprouted at a point below the graft, from the rootstock itself. The Putnams had grown the rootstock from seed. Gillett gave this tree to his son, who planted it in a field on the banks of the Ohio River.

The tree began to produce large, attractive red apples, and it was named Rome for the township in which it grew and Beauty for the attractiveness of its fruit. By 1828, it was being used for propagation. The original tree stood on the banks of the Ohio until about 1860, when it was washed away in a flood.

The Rome is a large, round apple, often an intense red. The flesh is yellow, sometimes a little mealy, and mildly flavored. It is a passable cooking apple, but overall it is mediocre. The fruit is common in the market, however, because the tree has several characteristics that endear it to the commercial grower.

Rome trees bear when moderately young and are productive almost to a fault. They are late blossomers, thereby avoiding frost damage, and are good pollinators of other varieties. The fruits are firm and stay on the trees till picked, so little of the crop is lost to windfalls. Romes can withstand rather rough handling and keep well. Finally, the large and regularly formed fruits are particularly suited to the machines used by commercial processers.

All these factors have combined to give a second-rate apple a very high billing. Trees are available from most nurseries.

Roxbury Russet

Although there are other contenders for the title, Roxbury Russet may be the oldest named variety of apple in America. In volume I of *The Apples of New York State*, S. A. Beach wrote that the Roxbury Russet was supposed to have originated in Roxbury, Massachusetts, early in the seventeenth century and that it was taken to Connecticut sometime after 1649. Since a seed-grown apple tree takes up to fifteen years to reach bearing age, the Roxbury Russet could be as old or older than Blaxton's Yellow Sweeting, which grew from a seed planted no earlier than 1635.

However, the descendants of a man named Joseph Warren claim that their ancestor grew the first Roxbury Russet and that it was at a much later date. The Warren family published a genealogy in 1854, and it records that Joseph Warren was born in Roxbury around 1696 and died there of a broken neck in 1755, when he fell from a ladder while picking apples. Further, the genealogy states, "He produced a russet apple with a red blush, called Warren Russet or Roxbury Russet." If the Warren family record is correct, then the variety is probably a century younger than Beach would have it.

The Roxbury Russet first appeared in Roxbury, Massachusetts, during the seventeenth century and is regarded as one of America's oldest named varieties. It is seldom found in cultivation today, despite immense popularity during the eighteenth and nineteenth centuries. From a lithograph published by G. & W. Endicott, New York, 1851.

Although the date of the origin of the Roxbury Russet is therefore in question, we do know it had spread throughout the Northeast by the end of the eighteenth century. The Putnam brothers (of Rome Beauty fame) took scions from their uncle's farm and introduced Roxbury Russet into the Midwest from their nursery in Ohio. In 1850, this variety (along with Rhode Island Greening and Winesap) was shipped around Cape Horn and planted in the Napa Valley of California. Through much of the nineteenth century it remained extremely popular.

The fruit was medium sized, round but slightly flattened, and just a little lopsided. The flesh was greenish white and juicy; the flavor was tart-sweet and very sprightly. It was excellent for cooking and for dessert, and for keeping it was second only to the Newtown Pippin.

The skin of the Roxbury Russet was its undoing, however. It was rough and mottled with yellow brown spots. Like other russet apples, it bore the nickname Leather Coat. So although the Roxbury Russet was a good-tasting apple, one of the best, and was a good keeper, lasting the entire winter in cellar storage, it wasn't very nice to look it, and as the twentieth century progressed and Americans abandoned growing and storing their own apples and began buying apples in the market, the Roxbury and all the russet clan slipped into undeserved obscurity. Refrigeration gave the edge to varieties with greater appeal to the eye.

The Roxbury Russet tree was hardy, vigorous, and very productive. Trees, at least in dwarf form, are available from Mr. Robert Nitschke, Southmeadow Fruit Gardens, 2363 Tilbury Place, Birmingham, Michigan 48009.

Stayman Winesap

"The Stayman Winesap originated as a seedling of Winesap grown by Mr. J. Stayman of Leavenworth, Kansas, in 1866," wrote S. A. Beach in *The Apples of New York State*, Volume I. It has been planted extensively in the Shenandoah Valley of Virginia and is grown principally in the Appalachian region. The tree amounts to an improved variety of Winesap, which it closely resembles.

The fruits are medium sized and bright red. The flesh is yellowish and crisp; the flavor is rich and aromatic. The Winesap is good for cooking, better for dessert, and is a good keeper, ripening in October. The trees are only moderately resistant to cold and are not recommended for northern states. Trees are available from most nurseries.

Wealthy

Wealthy was the achievement of Peter Gideon, the first American to breed apples scientifically, his chosen field for forty-one years.

Gideon was born in Woodstock, Ohio, in 1820, and in 1858 he settled on a land claim near Excelsior, Minnesota, on the shore of Lake Minnetonka. There he began developing fruits that could withstand the rigors of the northern climate. In 1859 he crossbred apples with the Siberian crab apple and planted the seeds the following spring. Out of the thousands of seedlings produced in that cross, several showed promise, and Gideon named the best of the lot after his wife, the former Wealthy Hull.

The fruit is large, bright red, and vaguely striped. The flesh is white, tinged with red, and the flavor is agreeably tart and rather aromatic. This apple is good for dessert or for cooking but is only a fair keeper. It ripens in October and should keep till Christmas, but not much after.

Wealthy trees are exceptionally resistant to cold and are best grown in the north-central states. The trees are vigorous, bear early, and will produce every other year, unless fruit is thinned in years with heavy crops, in which case it will bear fruit annually. Trees are available from many nurseries.

Winesap

William Coxe wrote in 1817 that Winesap was then becoming the most popular cider apple in West Jersey, an area noted for its cider. Little else is known about this variety's earliest history, though U. P. Hedrick opined that it probably was in cultivation in Virginia during the colonial period. By the mid-nineteenth century, Winesaps could be found throughout the southeast and middle states and had been introduced into California. Until about 1950, when planting and production began to drop, Winesap was one of the most popular varieties in America.

Winesap apples are medium sized and somewhat oblong in shape. The skin is deep red, with some yellow showing on the side of the apple that grew in shade. The flesh is yellowish, firm, and crisp; the flavor is sweet and aromatic. This apple is good for cooking or for eating fresh. It ripens in November and will keep until May. The trees are vigorous, only moderately resistant to cold, and are good producers. They are unusual in that they have pink blossoms instead of white. Winesap trees are available from some nurseries, but Stayman Winesap is offered more frequently

---

Yellow Transparent

This variety was imported from Russia sometime in the nineteenth century. It is an early apple that ripens in late July and early August and will keep into September. The fruit is pale yellow to almost white, with

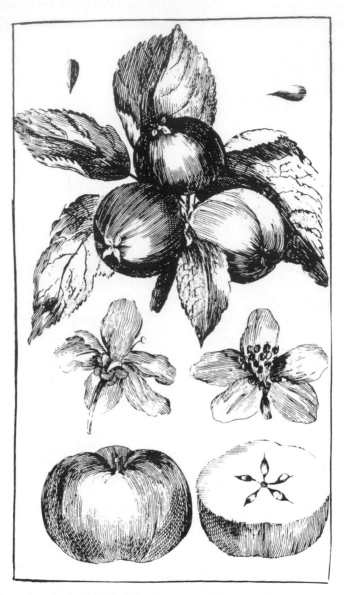

The pale yellow Calville Blanche was and is one of the celebrated cooking apples of France. This eighteenth-century bookplate shows the blossom from above and below, seeds, fruit, and a cross section of the fruit. From Henri Louis Duhamel Du Monceau's *Traité des Arbres Fruitiers*, Paris, 1782. (*Courtesy the Library of the New York Botanical Garden*)

a translucent skin. The flavor is tart, pleasant, but not terribly rich. It is an excellent cooking apple, but only fair for eating fresh. The trees are vigorous, produce abundantly, and are very resistant to cold. Yellow Transparent was one of the parents of several cold-resistant apples— Early McIntosh, Lodi, and Milton, all introduced by the New York Agricultural Experiment Station. Yellow Transparent trees are available from most nurseries.

York Imperial

York Imperial originated as a chance seedling on a farm near York, Pennsylvania. The owner of the farm, a Mr. Johnson, was first attracted to the tree one spring, when he noticed some boys digging in the leaves beneath it, searching for apples that had lain on the ground since autumn. He investigated and found the apples in good condition.

The next autumn, he brought samples of the fruit to a nurseryman, who began propagating the variety under the name of Johnson's Fine Winter. The trees did not sell, and the nurseryman dumped a lot of them in a roadside hollow. Passing farmers picked up the trees, took them home, and planted them. Within a few years, the value of the variety became clear, and Johnson's Fine Winter became quite popular in Pennsylvania and in several nearby states. The name York Imperial was suggested by Charles Downing, brother of A. J. Downing, during the 1850s. Downing considered the fruit an "imperial" keeper and joined that with the name of its hometown.

During the 1900s, many York Imperials were exported to England, and this export business spurred its popularity throughout the Appalachian region. The export business died about 1930 when Britain put rigid restrictions on apple imports, but the apple-processing industry stepped in and took its place. Today York Imperial accounts for about 5 percent of the total American apple crop, yet, because processors buy such a large share of the production, it is seen infrequently in the market.

The fruit of this variety is large and lopsided; the skin shows an unattractive mixture of red and green. The flesh is yellowish, crisp, and juicy; the flavor is mild but aromatic. It is a good cooking apple and quite acceptable for eating out of hand. York Imperial ripens in November and keeps through the winter. The trees are only moderately vigorous, so they require less pruning than many varieties. However, they produce heavily and will start bearing in alternate years if the fruits are not thinned in years with heavy crops. Trees are not commonly available at retail nurseries.

Purchasing Old-time Varieties

While many nurseries across the country specialize in the sale of old-time varieties of apple trees, the following offer the most extensive selection:

J. E. Miller Nurseries
5060 West Lake Road
Canandaigua, New York 11424

Stark Brothers Nurseries
Louisiana, Missouri 63353

Robert Nitschke
Southmeadow Fruit Gardens
2363 Tilbury Place
Birmingham, Michigan 48009

3

Growing Your Own Apples

Nineteenth-century bookplate illustrating a branch end with buds, a mature apple, and leaves and blossoms. From J. F. Schreiber's *Bilder-Werke für den Anschauungs-Unterricht in Schule und Haus*, Esslingen, 1882.

Growing apples doesn't take much. All you need is some space, lots of sunshine, and an apple tree.

The first thing to do is to look over your property to see whether an apple tree could live there. Apple trees aren't terribly fussy, but there are a few things they cannot tolerate. For example, if your land is better described as a swamp, maybe you should grow willow trees instead. They will do fine, but apple trees die if their roots are constantly wet. If your yard is surrounded by buildings or lots of trees, take a very careful look. Apple trees will not bear fruit unless they get full sunshine throughout a good portion of the day.

Finally, what are your winters like? Apple trees need a little cold each winter so they can rest. But too much winter can be too much of a good thing. If temperatures are dropping constantly to below 0 degrees F. and staying there, you may have problems. Beyond that, just avoid planting apples on windy hilltops or in little valleys or hollows. Strong winds can break branches and tear off fruit. Cold air can collect in hollows on spring nights and can frost apple blossoms and kill them, even when days are quite warm.

Choosing Your Tree

Choosing the right apple tree involves about half a dozen factors, and unfortunately, it is hard to say which should be considered first. But if you are going to put time and effort into planting and caring for an apple tree, you ought to choose the right tree in the first place.

Obviously, you will want to choose a variety of apple you like to eat, but that leads to a second question: eat fresh or eat cooked? Some apples are good for one or the other, and some are good for both.

When you have made that choice, another question comes up: Will the variety you prefer grow well in your part of the country? Some varieties grow better in certain areas, and if your choice is suited to your area, then what will you do with all the apples a good tree can produce? Some apples will keep for months in a cool cellar; others require cold storage if they are to keep for more than a few weeks.

What to do with all the apples leads to another point. Almost every variety of apple tree comes in several sizes, and each size produces fruit accordingly. Moreover, each size requires a different amount of space and a different kind of care.

Clearly, when there are so many questions, a sensible answer can only be found by isolating the questions and discussing them one by one. When you have all the individual answers you can find your own personal answer. There just isn't one answer for everybody.

CHOOSING A VARIETY

If you like several apples, it is probably best to narrow the field by eliminating those that serve only one purpose or are poor keepers. For example, you may be better off buying single-purpose Delicious apples at the market and growing all-purpose Golden Delicious. Otherwise, you will be either giving and throwing away a lot of apples or spending many hours canning applesauce (which isn't such a terrible fate). It is up to you to decide.

Then check to see whether your choice or choices will grow well in your area. The United States Department of Agriculture has issued a list that matches the right tree to the right area of the country.

The varieties recommended for northern states are McIntosh, Northern Spy, Rhode Island Greening, and Wealthy. For middle and southern states, Delicious, Golden Delicious, Stayman Winesap, and Jonathan are advised. Lodi, Yellow Transparent, Early McIntosh, and Gravenstein are recommended for all areas.

The USDA list is hardly exhaustive, because there are at least a hundred varieties on the market today. Of course, no single nursery carries them all. If none of the varieties mentioned matches your tastes and desires, the best course may be to check with your county agricultural agent or a reliable nurseryman.

Before making a final decision, there is one thing more you should do: Find out if there are any apple trees growing in your neighborhood. If there aren't, you may have to choose and plant a second variety, so the two trees can pollinate each other. Most apple trees will produce little or no fruit unless their blossoms have been cross-pollinated by a different variety. An explanation of this is contained in Chapter 2 in the section "An Apple Tree and How It Grows."

Golden Delicious and Rome Beauty will bear tolerable crops without cross-pollination, but they are exceptions. Rhode Island Greening and Gravenstein are also exceptions to the two-tree rule, but for the wrong reason. The Greening and Gravenstein are such poor pollinators that they cannot pollinate themselves or any other trees. Each should be planted near two other trees of different varieties. That way, the other trees can pollinate each other as well as the Greening or the Gravenstein.

Choosing a Size

Before you can decide which size apple tree to choose, you have to know what sizes are available. It would also help to know why the sizes are what they are and what can be expected of each, such as when they will start to bear fruit and how much fruit they will produce.

Generally speaking, each variety of apple tree comes in three sizes—dwarf, semidwarf, and standard. The size of the tree is determined by the kind of roots it has. Modern apple trees are not grown on what would be their natural roots. Instead, modern trees are two-piece trees, with the part that produces the apples grafted onto the part that produces the roots.

Some roots prevent the grafted upper tree from growing to full size. Depending on the kind of root, the tree may grow into a dwarf or a semidwarf. Other roots allow the upper tree to reach its natural size, and these become the standard trees. Just how big a dwarf or semidwarf will grow varies a bit, because different nurseries use different kinds of roots, and one cannot hope to describe every one of them. The dwarfs described below are trees grafted onto root varieties classified by the East Malling Fruit Research Center in Kent, England. These roots, or rootstocks, as they are called by nurserymen, are commonly used in America, and the name of the rootstock used to produce a particular dwarf is often available from a local nursery.

The smallest dwarfs usually grow on a rootstock called Malling IX, often abbreviated as EM IX. They will grow to a height of only six to nine feet in twenty years. They should start to bear fruit within three years of planting, just a few fruits at first, and can be expected to produce two or more bushels of apples annually as they approach full growth.

Semidwarf trees can grow from two rootstocks, Malling VII (EM VII) and Malling II (EM II). Malling VII trees grow twelve to fifteen feet tall, bear after two or three years, and can produce four or five bushels of fruit. Malling II trees grow to twenty feet, bear a little later, and can produce eight or more bushels of fruit. The standard tree will usually grow to thirty or forty feet, will start bearing after five years, and can produce from fifteen to thirty bushels of fruit.

Unfortunately, none of these size and yield estimates can be regarded as 100 percent accurate. The final size of the tree depends on the variety of apple as much as it does on the type of rootstock. Also, the amount of fruit each tree produces depends on the variety, the fertility of the soil, and the general health of the tree.

Now it is time to decide which size is best for your purposes, and a quick way to narrow the choices is to consider whether you will care for the tree or trees yourself or hire someone else to do it.

First off, every apple tree needs a certain amount of pruning. Certain branches have to be cut back each year to keep the tree from choking itself with so many branches that fruit begins to suffer because of it. Pruning a dwarf requires little equipment and offers little hazard to the amateur because tall ladders and climbing aren't involved. A semidwarf takes more equipment, and the work is more dangerous. The same applies to the standard tree, only more so.

Every fruit tree also requires a certain amount of spraying each year to protect it from insects and diseases. A dwarf can be sprayed with hand-operated equipment. Full-grown semidwarfs and standards are too big to be sprayed with anything but power equipment. Tree experts or tree surgeons can be hired in most parts of the country to do the jobs that are literally out of the homeowner's reach, but tree experts charge for their services. The choice is up to you, but if you still can't decide, there is also the question of space.

A dwarf tree may grow nicely in the middle of an area measuring twenty by twenty feet. A semidwarf may require forty by forty feet to grow just as nicely. A standard tree requires proportionately more space. But remember: All the space doesn't have to be on your land. Trees don't know anything about property lines. It doesn't bother a tree one bit if the sunlight it needs passes over your neighbor's land first. The real criterion as far as space is concerned is whether the tree will get enough sunlight.

Remember, too, that if you have to plant two or more trees for cross-pollination, two semidwarfs will fit in the space required by a single standard tree and four dwarfs will grow in that same space. Now there are two or three more factors that can help you make the choice of size.

Standard-size trees are generally healthier and longer-lived than dwarfs or semidwarfs. If space is no problem and if you can care for the tree yourself or are willing to hire a professional, then a standard tree may be your best choice.

Semidwarf trees are less rugged than standards and pose many of the same problems when it comes to pruning and spraying. They are a good compromise if space is a consideration.

Dwarfs are the least rugged and long-lived. Their roots are so weak that the trees must be anchored in place with stakes and guy wires. The part of the wire that loops around the tree should be covered with a length of rubber garden hose or something like it, and the whole apparatus is a bit unsightly. However, in tight spaces dwarfs are the best, and they are easy to care for.

In the very tightest spots, dwarf trees can be trained to trellises, with all but a few branches cut away. These are called espalier trees. They can be purchased already trained, or the grower can train them himself. Trained or untrained, dwarfs can even be grown in big wooden tubs, which makes it possible to grow apples on a city rooftop or in a swamp. Information about both of these techniques—espalier and tub-grown trees—can be found in any number of books dealing specifically with dwarf trees and are available in any good library.

The New Arrival

Let's say you have ordered your tree and parcel post has left it on your doorstep. What next? Assuming that you have purchased your tree from a dependable nursery, the answer is to plant the tree as quickly as possible. A good nursery will send your tree out so that it will arrive during the proper month for planting. This is March or early April in

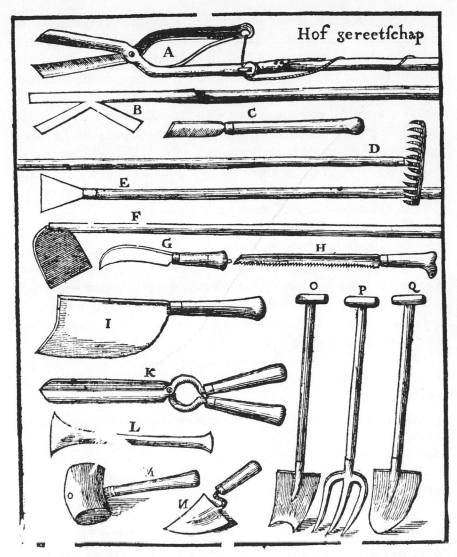

These implements were among the orchardist's arsenal of tools in the seventeenth century. From Jan van der Groen's *Der Nederlandtsen Hovenier*, Amsterdam, 1683. (*Courtesy the Library of the New York Botanical Garden*)

areas where winters are severe. In other areas it is late October or November. Whether spring or fall, the ground should be warm and dry.

Dig a hole at the spot you have chosen and dig it round enough to hold all the tree's roots without crowding them. As you dig, first put the topsoil to one side of the hole, then the subsoil to the other. (The subsoil is usually a different color.) Dig the hole deep enough so the tree can be planted at the same depth it was at the nursery. In dry areas, a tree may be planted a little deeper to keep the roots from drying out in summer heat. In damp areas the tree can be planted just a little shallower, so the roots won't be constantly wet. Planting too deep is a more common error than planting too shallow.

Now put some of the topsoil into the bottom of the hole and spread the tree's roots over it. Cover the roots with more topsoil and firmly pack the earth by trampling it underfoot. If you can get compost or well-rotted manure, mix it half and half with the remaining topsoil and the subsoil. If the topsoil is very black or smells sour, a pound or two of lime can also be mixed with the remaining soil at this time. The lime neutralizes soil acids.

On no account should fresh manure, dry manure powders, or any chemical fertilizer be used at planting time. Strong fertilizers can chemically burn the roots of an apple tree. Furthermore, if nutrients are too readily available, the young tree will put forth lots of new branches but will not develop a good enough root system to sustain those branches through the first dry spell.

The topsoil right around the roots has to be firmly packed to anchor the tree in place. Now fill the rest of the hole loosely, so water and air can get down to the roots. Apple tree roots need both air and water to grow properly. Now there is one more step required to finish the job of planting, but there are two ways of doing it, depending on the season.

If you are planting the tree in autumn, give the tree a good watering, then mound the earth into a little cone around the base of the tree. This will prevent water from collecting around the roots and making the soil heave up and down as the water freezes and expands, then melts and contracts. This heaving can tear the slender roots of a young tree. If you are planting the tree in spring, water the soil, then make a little depression, about two feet in diameter, around the base of the tree. This will collect rain in summer and funnel it to the roots. If you planted the tree in autumn, give it the same rain catcher in spring. In either case, don't let this little depression become a lake during a rainy spell; the tree might drown in it.

Care After Planting

If you have planted your tree properly, it will need little further care during its first summer. Just protect it from the usual hazards of tree life. Keep the area around the trunk free from weeds and rubbish. Dead leaves, weeds, and debris can provide hiding places for mice, which like to feed on bark, and breeding places for insects that can attack any part of the tree. If you find that insects are chewing holes in the leaves or if you notice tiny insects, called aphids, clustered on the tender branch ends, you should apply an insecticide such as Malathion, diluted according to the manufacturer's instructions. Don't assume that you know more than the manufacturer by making the mixture stronger than recommended. You are very likely to injure the tree that way. Malathion is a very useful general purpose insecticide that effectively controls many types of insects but is cursed with a rather obnoxious odor. Fortunately, the odor disappears when the spray dries.

If you have no spray equipment, don't run out to buy some huge, elaborate tank model at this time, though you may want one later. With the tree as small as it is, you won't need much spray material, and you can't let the mixture lie in the tank for weeks without damaging the sprayer. A moderately priced, hand-held model, with a one-quart capacity and a continuous spray will hold you in good stead for the next few years. (The atomizer-type sprayer, which sprays only when you push the handle, is less expensive but not nearly as effective as the continuous-spray type. Too much foliage is left unsprayed with the atomizer type.) Spray the tree when you notice insect damage or when you see the insects themselves. There is no need to follow any sort of spray schedule at this time. That will come later, when the tree starts to bear fruit.

There are a few other points of care for the year-old tree. If the weather is very dry, give the roots a thorough watering once a week. If a branch gets broken, cut the broken part away so that the cut is clean and won't collect water. If the tree starts to lean to one side, drive a stake into the ground on the opposite side, far enough away so you don't hit the roots. Then pull the tree up straight with some loops of soft rag. These won't injure the bark. A tree cannot be straightened up when it grows big, and a leaning tree is far more likely to topple in a winter storm in later years.

The Second Year

This is the year you start giving the tree the kind of special care that will help it bear lots of fruit for many years. You begin by helping it develop a set of strong branches. Early in the spring, before the tree has started to grow again, get yourself a good pair of pruning shears, if you don't own them already, and pay your tree a visit.

During the previous summer the tree should have produced side branches all along its trunk. Now starting at the top of the tree, look for any branches growing at less than a 45-degree angle from the trunk. These branches are trouble. Should one of them break later on, half the trunk could go with it. Cut these off now. (See the section on "Pruning Your Tree," page 80, for further instructions.)

If there are two branches at the very top of the tree that make the tree look like the letter Y, cut off one of the branches. The idea is to give the tree a strong central leader.

If there are any branches lower than two feet from the ground, cut them off, too. These low-down branches will eventually droop to the ground and will make it impossible to clean up the area around the trunk.

Just be careful of one thing: If a low-down branch is nearly as thick as the trunk, which is often the case, cutting it off next to the trunk could cost the tree as much as a third of the bark in that area. This could kill the tree, because the bark is more than just the skin of a tree. The bark is divided into two layers. The outer layer is dead and dry, but the inner layer is green and alive. Actually, it is the only living part in the trunk or branches of a tree.

The inner bark produces both the wood, which also is dead, and the outer bark. It is the inner bark that carries water and soil nutrients from the roots to the leaves and returns the sugar that the leaves produce from water, air, and sunshine to the roots, which use the sugar for food.

That is why you can't deprive a tree of a third of its bark, even in one area, without risking fatal consequences. So if the low branch you intend to cut is too thick, cut it back to a six-inch stub. The tree will keep growing, and in a year or two you can cut the stub flush with the trunk. By then, the loss will represent only a tiny fraction of the bark encircling the trunk. If any shoots sprout from the stub, cut them off, or the stub will grow thicker, defeating your whole purpose.

During the second year you should give the tree its first chemical fertilizer. Use the mixture called 5–10–5, which contains 5 percent nitrogen, 10 percent phosphorus, and 5 percent potassium. These are elements essential for tree growth. In late March or early April scatter one pound of 5–10–5 over the area that is within three feet of the trunk. Scatter it evenly, breaking the surface of the ground by lightly scratching it with a rakelike tool, or cultivator, and the rain will carry the chemicals down to the roots.

After pruning away undesirable branches and applying the fertilizer, no further care should be required beyond the general maintenance already described, such as keeping the trunk area clean, spraying for insects, and watering in case of drought. However, should the tree blossom and start to produce fruit, as a dwarf may very well do, pick off all but a couple of the apples while they are still little nubs. Give the tree another year to gain strength without the added burden of a heavy crop of fruit. A tree overburdened at an early age may never regain its strength and may bear poorly for the rest of its life.

The Third and Fourth Years

During these two years, the goal once more is to provide the tree with a strong set of branches. The rules are the same as they were in the second year: Cut off any branch growing at less than a 45-degree angle from the trunk and don't allow the tree to form a second leader at the top. Don't prune any other branch unless it is damaged. There is an old European tradition of pruning the Dickens out of young trees to channel their energies into fruit production. This method doesn't work in America because growing conditions are different. An American tree will produce more fruit sooner if it hasn't been pruned heavily in its early years.

CONVENIENT PORTABLE FRUIT-
LADDER AND STAGING
From S. E. Todd's *The Apple
Culturist*, New York, 1871.
*(Courtesy the Library of the New
York Botanical Garden)*

You also should give the tree more 5–10–5 fertilizer—three pounds in the third year and four pounds in the fourth year. The rule of thumb is to match the number of pounds to the age of the tree. However, you can give an apple tree too much fertilizer, and if you do, the tree will put all its energies into growing leaves and branches. Fruit production will drop, so you have to strike a balance between too little fertilizer and too much, and there is a way to do it.

In the third year put down one and a half pounds of fertilizer in late March or early April, then wait till the end of July and look at the ends of the branches. When an apple tree is in good health, the ends of the branches will grow about a foot in length between April and the end of July. Sometimes it's a little more, sometimes it's a little less. Look at the ends of the branches. The new growth has smooth bark, and there is a definite kind of joint where the new growth starts. If the branches have grown only three or four inches since April, apply the other one and a half pounds of fertilizer. In the fourth year, give the tree two pounds of fertilizer in late March or early April, then wait till July and check the branch ends. If growth appears normal, hold back the fertilizer; if not, apply it.

If you planted a dwarf tree, fruit should be part of the picture in the third and fourth years, especially if the tree bore fruit during the second year. This means you will have to start spraying the tree systematically to protect it from insects and disease. The spray season begins as soon as the buds show a half inch of green leaf tissue. (The section on "Spraying Your Tree" outlines a typical spray schedule and describes the materials used.) If the tree fails to blossom or to "set" fruit, you can return to spraying the tree only when insects or insect damage are noticed. However, you must resume the schedule the following spring on the assumption that the tree will begin to bear fruit.

If the tree does set fruit, you may notice in late May or early June that a number of tiny apples have fallen from it. This is nothing to be alarmed about. What happened is that the tree set too many fruits back in early May and now is shedding the excess. However, even after the so-called June drop, the tree may still be carrying more fruit than it can handle. Most apple trees habitually overbear, and if the fruit is not thinned by hand, the apples will be small and poor in quality at harvest time.

A week or so after the June drop is over, study the tree carefully. If you find apples growing in clusters of two or three, they should be thinned to a single fruit. Next, look along the length of each branch to make sure there is at least six inches between one apple and the next, and pluck off the offenders. In some years, no thinning will be necessary, or else it will be needed on one or two branches only.

The Fifth to Twelfth Years

During the fifth through twelfth years, you begin to select the branches that are called the main scaffold of the tree. These are three to five boughs springing from the trunk that will carry (along with the top growth of the central leader) all the mature tree's fruit-bearing branches.

The purpose of choosing the main scaffold and emphasizing its growth is to keep the area around the trunk as open and free from branches as possible. This allows air to circulate around the trunk and inner branches, keeping them dry. It also allows a certain amount of sunlight to filter through to the center of the tree. This keeps the center of the tree from becoming the breeding ground for fungus and other diseases that thrive in moist, dark places.

Here are the criteria for selecting the main scaffold branches: Choose three to five branches that are well spaced around the tree, each pointing toward a different point on the compass. For example, choose one branch that grows northward, one that grows eastward, and so on.

PLUCKING APPLES AND PUTTING THEM IN A GRAIN-BAG HUNG OVER ONE SHOULDER
From S. E. Todd's *The Apple Culturist*, New York, 1871.

Choose branches that are at slightly different heights on the trunk, if possible. The branches chosen should be horizontal to the trunk or growing slightly upward. Downward-growing branches or those growing upward at an acute angle from the trunk should be marked for elimination. However, cut off only one of the undesirable branches during the fifth year, then a second undesirable branch in the sixth year, then one more branch each year thereafter until only the branches you have chosen are left. Cutting off all the objectionable branches in a single year could kill the tree. Moreover, if some of the branches you have chosen for the scaffold get broken, you still have other branches springing from the trunk that can be chosen as substitutes.

As the years go by, some of the branches will grow so thick that pruning shears won't cut through them. Then you will have to use a pruning saw. (A description on how to use this tool is included in the section on "Pruning Your Tree," page 82). Furthermore, when you have removed an undesirable or damaged branch, you may leave behind a wound big enough to require special tree paint, which is available from most farm and garden suppliers. Tree paint contains an antiseptic to keep the wound from becoming infected by plant diseases. The paint also contains ingredients that hasten healing and should be applied to any wound larger than one and a half inches in diameter. Smaller wounds heal quicker if left alone.

It is just about impossible to give Delicious or Rome Beauty trees an open center by the main-scaffold method. Delicious produces most of its branches at an acute upward angle from the trunk, while Rome Beauty produces branches in tight whorls. If you grow either of these varieties, just try to keep the center of the tree as open as possible by cutting out weaker branches and leaving the stronger.

During the fifth through twelfth years you will also start cutting back the ends of the main branches to give the outside of the tree the outline of a pyramid or cone. If the tree is allowed to grow freely, it will start to take on an umbrella shape. The upper branches will keep sunlight from reaching the lower branches, and the lower branches will stop bearing fruit. However, if the tree is shaped like a pyramid or cone, all the branches will get equal sunlight and all will produce fruit.

As the tree begins to age, it will begin producing light green sprouts along the lengths of the branches and on the trunk. These sprouts are called by a variety of names, such as water sprouts, suckers, or risers. In most cases, they should be cut off flush with the bark. These sprouts will produce little or no fruit themselves and will choke out the branches that do produce fruit. There are two cases in which a sucker may be allowed to remain. First, if a limb has to be removed because of disease or damage, a sucker can be allowed to grow and replace the missing branch. Secondly, when a fruit branch bends downward, it often will stop bearing fruit. If a sucker happens to appear at the point where the branch bends downward, you can cut off the drooping part and let the sucker replace that.

Other care during the fifth through twelfth years is essentially the same as for the first four years. Keep the area around the trunk free of weeds and rubbish. Now it is advisable to "mulch" the area under the branches with a three- to four-inch-thick layer of hay, straw, or even grass clippings. The mulch will keep down the weeds and will prevent moisture from evaporating out of the soil. Keep the area immediately around the trunk free of mulch to prevent mouse and insect damage. The mulch should cover an area of three to four feet out from the base of the tree, and as the branches grow, you should put down more mulch, so the ground is covered just beyond the spread of the branches.

The yearly applications of 5–10–5 fertilizer should also continue, matching the number of pounds to the age of the tree. As before, only half should be applied in late March or early April and the rest withheld until the end of July, when you should check the branch ends for new growth, as described before. If the new growth measures only three or four inches, give the tree the rest of the fertilizer. After ten years, however, you should divide the fertilizer differently. The number of pounds should still match the tree's age, but apply only five pounds in the spring, holding back the rest until the end of July, then applying or withholding the remainder according to the amount of new growth.

Naturally, the full spray schedule still applies to dwarf trees, and now the same begins for standard trees. As with the three- to four-year-old dwarf tree, you should start spraying the standard tree each spring when the buds show a half inch of green. Begin with the standard spray schedule, then, if the tree fails to blossom and set fruit, return to spraying only when insects or insect damage are present. The standard tree should start bearing between the fifth and tenth years. In years when the apple crop is heavy, the fruit on both dwarf and standard trees must be thinned to eliminate clusters of two or three apples and to make sure there is at least six inches between apples along the length of any branch.

The rest remains the same as before: Water the tree in the event of drought. An easy way to do this is to run a garden hose out to the base of the tree and barely open the valve, so the water just trickles out. Let the water trickle into the ground overnight, doing this about once a week for as long as the drought continues. If a branch breaks, cut it off clean so water won't collect in the injury, making the spot a breeding place for plant diseases. As mentioned before, if the wound is bigger than one and a half inches in diameter, apply tree paint. Finally, at the risk of sounding repetitious, keep the area around the tree as clean as possible. Dispose of fallen fruit quickly and effectively. Often an apple falls from a tree because there is an insect within gnawing at its innards. If you leave the apple lying around, that insect will live to spawn hundreds more. There is no point in breeding your own pests.

The Thirteenth Year and Onward

Asking how long an apple tree will live is something like asking "How high is up?" A dwarf or semidwarf tree (their care and characteristics are similar) can be expected to bear well until sometime between its twentieth and thirtieth years, assuming it has received good care and hasn't been damaged by disease or mechanical injuries. It can live beyond 30 years, but fruit quality will diminish as will the size of the crop. A standard tree should show little decline in fruit quality and crop size until well into its fourth decade, and it could survive for a century or more. There are cases on record where apple trees of known varieties were still bearing some fruit after 150 years. In most cases, you can expect to be rewarded if you give your tree good care for at least another 15 to 30 years. The care is much the same as it has been from the second year on.

The main scaffold should have been established by the twelfth year, and pruning now is aimed at keeping the tree from choking up with branches and suckers and keeping it in the shape of a pyramid. However, as a tree continues to age, pruning takes on a third importance.

As an apple tree ages, its branches begin to sag, and this sagging seems to be some kind of signal to the tree to go easy on fruit production. The more the branches sag, the less fruit they produce, but this can be corrected to a great extent by pruning. Whenever a sagging branch stops bearing fruit, it can be cut off at the point where the sag begins. Usually, the tree will throw out a sprout, just behind the point where the cut was made. If the sprout grows upward, which is likely, the odds are it will start to bear fruit in a year or two.

We have now established the three reasons for the importance of pruning during the entire productive life of the tree. First, by thinning suckers and excessive growth you keep the center of the tree open to light and air. This helps in controlling diseases. Second, by keeping the tree in the shape of a pyramid, you guarantee that every branch will get enough sunlight for fruit production. Third, by removing sagging branch ends and inducing them to grow upward, you keep the tree "thinking young" and encourage it to produce new fruit-bearing wood.

Of course, older trees need everything they needed in their earlier years: Keep the area under the branches clean and well mulched. Keep mulch, weeds, and debris away from the trunk. Remove damaged branches with a clean cut and apply tree paint as needed. Give the tree fertilizer according to the pound-per-year formula. Spray the tree yearly, following your spray schedule, and water the tree weekly in the event of drought. If you do all these things, you can expect fine crops of apples year after year for decades and all the pleasure that goes with doing something right.

Pruning Your Tree

Pruning requires few tools, but they must be in the best condition. The only tools needed for a dwarf tree are a pair of pruning shears, a pruning saw, and a small stepladder. For a standard tree, you can use the same tools, a larger stepladder, and a pair of rubber-soled shoes, so that you can climb into the tree without injuring the bark. But the standard-tree owner may want to add other tools to his arsenal: a pole pruner, which is a pair of pruning shears mounted on the end of a long pole and operated with a pull cord; and a pole saw, which is a curved saw similarly attached to a long pole.

PRUNING SHEARS

Pruning shears come in three patterns, but the best for general use is the double-cut type, which has two cutting edges. These will cut branches as thick as your fingers with the greatest ease. These shears

AN APPLE-TREE PROPERLY TRAINED AND PRUNED

Note that the tree has been shaped into a rough pyramid, which allows equal sunlight to reach each branch. The metal collar and the sticky bandage around the trunk as well as the two bottles used as traps were various nineteenth-century measures aimed at insect control. From S. E. Todd's *The Apple Culturist*, New York, 1871. (*Courtesy the Library of the New York Botanical Garden*)

AN APPLE-TREE WRONGLY TRAINED AND PRUNED

Note the umbrella shape of the tree and the spindly shoots, called suckers, springing straight up from the lower branches and surrounding the base of the trunk. From S. E. Todd's *The Apple Culturist*, New York, 1871. (*Courtesy the Library of the New York Botanical Garden*)

look something like a pair of scissors, but the proportions are reversed. The handles are longer than the blades. Furthermore, one blade is relatively thin and curved, while the other is thick and flared so it will fit the curved blade.

Whenever you use pruning shears, squeeze the handles firmly, without wiggling them. This way you get a clean cut. Ragged cuts are dangerous to the tree because water, fungus spores, and bacteria collect on the rough surface, making it a likely place for infection to start and spread throughout the tree. If you cannot make a given cut using pressure alone, get out your pruning saw.

When cutting small branches or suckers from the trunk or a larger branch, cut as close to the trunk or branch as possible. This way bark will grow over the wound quickly. If you leave stubs all over the tree, they can rot and become places at which infection can enter the tree. (There is that one exception mentioned before, where the very lowest branches of a two-year-old tree are left as stubs for a year or two. In that case, the risks involved in leaving the stubs are less than those involved in cutting them off.)

When cutting off suckers or small branches, also avoid cutting down through the crotch, the place at which the small branch joins the larger one. Put a blade on either side of the branch or sucker and squeeze the handles firmly. Again, this will provide the cleanest possible cut.

POLE PRUNER

This tool is especially useful for pruning the branch ends of tall trees. As mentioned before, a pole pruner is essentially a pair of pruning shears mounted on a pole, and the uses of this tool are about the same. Apply pressure to the pull cord firmly and smoothly; jerking the cord amounts to the same thing as wiggling and twisting a pair of shears. If you can't make the cut with even pressure, use a pole saw.

PRUNING SAW

Pruning saws are usually curved and are designed specifically for cutting through trees. A carpenter's saw is designed to cut through dry, "cured"

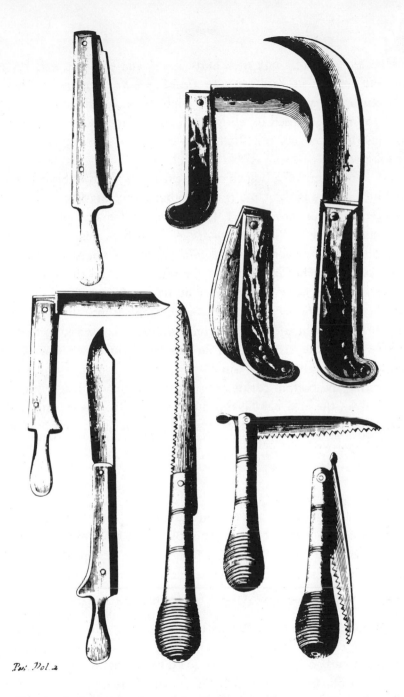

Peri. Vol. 2

Pruning knives and saws used in the seventeenth century. From Jean de la Quintinie's *The Compleat Gard'ner*, "made English" by John Evelyn, London, 1693. (*Courtesy the Library of the New York Botanical Garden*)

wood, and the blade will stick and jam if you try to use it on the relatively damp wood of a live tree. Pruning saws come in a variety of sizes, but a saw with a fourteen- to eighteen-inch-long blade will be ample for most purposes. Except when a large branch has been damaged and must be removed, you should never need a saw bigger than the one suggested.

When cutting off a branch near the trunk or another branch, try to avoid nicking the bark of the trunk or that larger branch. Once again, when removing an entire branch (rather than a damaged piece near the end), cut as close to the trunk or larger branch as possible. This way the bark will grow over the cut. Again, when the wound is more than one and a half inches in diameter, apply tree paint to the wound.

Finally, if a large branch must be removed because of disease or other damage, begin by sawing a two-inch-deep cut into the bottom side of the branch, about eight inches from its base. Next, working from the top side and about six inches from the base, cut down through the branch until it falls. If you simply try to cut from the top without first making the bottom cut described, the branch will tear a big piece of bark from the area behind the cut when it falls. Once it has fallen, saw off the stub as close to the trunk or branch as possible and dress the raw wood with tree paint.

POLE SAW

As the name implies, this is a pruning saw fastened to the end of a long pole. It is used to cut off branches too thick for a pole pruner. The pole saw is not suitable for use on really big branches, however. For those, a ladder and hand saw are best. You have much more control over a hand saw and can make far more accurate cuts.

PRUNING TIME

Pruning should be done during the tree's dormant period, after the leaves have fallen but before growth begins again. It is easier to see exactly what needs to be done when the leaves are gone and there are less leaves to disturb. Avoid freezing cold days. The cold can injure freshly cut plant tissue. Pick a warm, bright day; it's better for the tree and more pleasant for you. Damaged branches, of course, should be removed as soon as possible no matter what the season or temperature. Wind twisting a dangling branch can cause even more damage by tearing the bark below the point of fracture.

Spraying Your Tree

Knowing when to spray and what chemicals to use is actually more important than having detailed information about the bugs you're fighting. So let's establish a sample spray schedule first, then go on to describe the various insects and apple disorders. This schedule is designed for use in the Northeast, but it is typical of most spray schedules. To get a spray schedule for your specific area, contact your county agricultural agent.

A spray schedule doesn't start on any specific date; rather it begins when the apple tree starts showing signs of life. When the buds start to open and a half inch of green leaf tissue is showing, the tree is sprayed with a mixture of five tablespoons of miscible oil per gallon of water. This is called a half-inch spray and is aimed at killing scales and mites.

The second spray is applied just before the blossoms open, when they look very pink. The spray is called the full-pink spray and contains a fungicide and an insecticide. It is aimed at a disease called apple scab and at aphids. The most popular full-pink spray is a mixture of Captan (a fungicide), Malathion (an insecticide), and water.

The third spray is applied when 90 percent of the petals have fallen, and it is called the petal-fall spray. It contains the same fungicide and insecticide as the second spray, as well as a second insecticide, methoxychlor. It is aimed at apple scab and such insects as curculio, caterpillars, and mites.

The next two sprays are identical and are called the first and second cover sprays. The first cover is applied two weeks after "full-pink." The second cover is applied two weeks after the first. Each contains Captan and methoxychlor and is aimed at scab and such insects as codling moth and curculio.

The third cover spray is applied three weeks after the second cover spray. It contains the same ingredients as the second cover spray plus the addition of Malathion and is aimed at scab and fruit rot and such insects as codling moth, apple maggot, and aphids.

The fourth through sixth cover sprays are identical to the third in terms of composition and are applied to combat the same diseases and insects. They are applied at three-week intervals.

For a dwarf tree, a tank-type sprayer will be ample to apply all these sprays. The same will be true for a standard tree until it gets too big. Then you will have to invest in a power sprayer. However, in most areas you can hire a tree service to apply all of these sprays on a contract basis. One advantage here is that the tree service employees know exactly when to apply each spray.

When spraying fruit trees, remember that you can't apply too much spray. Cover the leaves, branches—everything—as best you can. The excess will simply drip off. It also is advisable to add what is called a spreader-sticker to each batch of spray. As its name implies, it helps the spray to spread and stick. As always, follow the manufacturer's instructions.

Apple Diseases

Apple Scab

Apple scab is a disease caused by a type of fungus. The fungus usually appears on the underside of the leaf as a fuzzy, dark green spot that ultimately turns black and looks rather like a scab. Fruit can also be attacked by this fungus. Following the spray schedule described will control this disease in fruit-bearing trees. Should it attack the leaves of a young fruit tree, you should apply two tablespoons of fungicide such as Captan, mixed with a gallon of water. Repeat the spray every three weeks through the rest of the summer. This disease is a common problem in humid areas.

FIRE BLIGHT

This disease is caused by a type of bacteria that blackens leaves and fruit and makes infected branch ends look as though they have been scorched. Some apple varieties, such as Jonathan, Wealthy, and Yellow Transparent, are particularly susceptible to this disease.

The blight usually appears in spring, when the trees are blossoming, and remains active for a month after blossoming. New growth is more susceptible to the disease than older wood, so during that summer, don't give the tree fertilizer, which will encourage new growth.

Cut off infected twigs as soon as they are noticed. Make the cut twelve inches below the infected part in summer and six inches below during the dormant season. After cutting one twig and before cutting the next, dip your pruning shears into a mixture of household chlorine bleach and water diluted to a strength of one part bleach to ten parts water. This will prevent your spreading the disease germs with your shears. The bleach kills the germs. When you have finished with the pruning shears, wash them well with water to remove all the bleach, which can make them corrode. Dispose of all blighted twigs, so they are not a source of contamination to the tree once more. Since aphids are suspected to be spreaders of this disease, make sure you control this problem as soon as it is noticed. Two tablespoons of Malathion diluted in a gallon of water is a common antidote. The regular sprays applied to fruit-bearing trees are also designed to control aphids.

POWDERY MILDEW

This disease is caused by a type of fungus that makes infected twigs and branches look as though they have been sprinkled with a white powder. This disease can be controlled by spraying with a mixture of three tablespoons of wettable sulfur powder to a gallon of water. This spray should be repeated every two to three weeks until midsummer, when the disease becomes inactive.

Other diseases affecting the apple include bitter rot, a problem in hot, moist areas, and black rot. These diseases principally affect the fruit and should be controlled by the chemicals used in the regular spray schedule for fruit-bearing trees.

Insect Pests

Although some thirty insects are known to prey on the apple tree, its leaves, and fruit, some are so rarely encountered that they really are not worth mentioning. The insects described in this section cause most of the grief. All should be controlled by the insecticides used in the spray schedule already described, but some people prefer not to use such a "shotgun" approach to the problem, so measures specific to each insect have been included. Insecticides have been named, but specific instructions on how to dilute them have not; simply follow the manufacturer's recommendations.

APHIDS

Aphids are tiny soft-bodied insects that vary in color. Those affecting apples are usually light green, dark green, or a purplish black. Some aphids have wings; others are wingless. Often they are found clustered on young, tender stems or on the undersides of leaves. These insects cause the leaves to curl and thicken, turn yellow, and ultimately die as they suck out the leaves' juices.

To control these pests, apply the insecticide Diazinon or Malathion whenever aphids are found in numbers.

APPLE MAGGOTS

Apple maggots are little wormlike creatures, yellowish white, and growing to a length of about three eighths of an inch. They are the immature forms of a black fly that has green eyes and white stripes on its abdomen. The adult fly lays its eggs on the apple.

Apple maggots burrow into the fruit, distort its shape, and cause it to fall prematurely from the tree. This pest is rare in southern states.

To control apple maggots, apply Diazinon or methoxychlor three or four times, ten days apart, when the adult flies appear, which is usually in late June or early July.

BAGWORMS

Bagworms are the larvae or immature forms of a variety of moth. The adult is rarely seen, so a description wouldn't be helpful here. The male of the species has wings and is free to fly about after dark. However, the female is legless and confined to the conspicuous, baglike case that she weaves for herself while a caterpillar. Young bagworms eat the leaves of apple trees. The species is only found east of the Rockies.

To control bagworms, apply the insecticide Carbaryl or Malathion in May or June, when the young bagworms appear. On a small tree it may be easier to simply remove the bags and burn them.

CANKERWORMS

Cankerworms, like bagworms, are the larvae of a type of moth whose male has wings but whose female is wingless. The male is an attractive gray moth with a wingspan of about one and a quarter inches. The female is a plump little bug about a half inch long.

The larvae, which cause the damage to trees, are the familiar inchworms or measuring worms, which are often seen dangling down from trees, each on a silken thread. The worms use these threads to get to the ground, where they pupate—that is, turn from caterpillars into the adult moths. The worms vary from light brown to dark brown but all have yellow stripes running down each of their sides.

To control cankerworms, apply Carbaryl when the buds show pink or as soon as the worms are noticed.

CODLING MOTHS

The codling moth was introduced from Europe and has become a major pest in America. Adults of both sexes are pretty little gray and brown

moths with wingspans of about three-quarters of an inch. The caterpillars are a creamy or pinkish white with brown heads. They grow to
about a half inch long. When you find a worm in an apple, most likely
it is a codling-moth caterpillar. The moths spend the winter as pupae
in cocoons and emerge as adult moths in the spring to lay their eggs on
apple leaves. When these eggs hatch, the caterpillars burrow into the
young fruit, causing little blemishes on the apple skin. Later generations,
and there can be several in a season, hatch from eggs laid directly on
the fruit.

Codling moths can be controlled by applying methoxychlor ten to
fourteen days after petals fall, then reapplying every ten days while fruit
is developing.

FALL WEBWORMS

The fall webworm is a handsome white moth with black or brown
spots on its forewings and a wingspan of an inch and a half. The caterpillar is a hairy worm—black with orange spots. Unlike the tent caterpillar, which usually spins its webs around crotches, the fall webworm
makes its webs at the ends of branches, covering a cluster of leaves.
These caterpillars feed on leaves and are capable of completely defoliat-
ing a tree.

The easiest way to control these pests is to take a long pole with
several nails driven into the end. Wind the webs around the end of the
pole, then burn them. Don't burn the webs while they are still in the
tree; you will damage the tree that way.

Japanese Beetles

The Japanese beetle is another pest that was introduced into America
from another part of the world. When the beetle first appeared in this
country, in New Jersey in 1916, experts were hard pressed to collect a
dozen specimens. Today, the beetles pose a major insect problem east of
the Mississippi.

Adult Japanese beetles are about a half-inch long, bright metallic green, with reddish brown outer wings. The adults feed on foliage, while the inch-long larvae and grubs, which are white with brownish heads, feed on the roots of grass and other plants.

To control adults, apply Carbaryl, methoxychlor, or Malathion to infested foliage. The insecticide chlordane is used to treat larvae-infested lawns, but Japanese beetle grubs can be controlled organically by applying milky-disease spores, which infect the grubs and kill them. For information on this method, send a postcard to the Office of Information, U.S. Department of Agriculture, Washington, D.C. 20250, and request Leaflet 500, "Milky Disease for the Control of Japanese Beetle Grubs."

Plum Curculio

The plum curculio is a tiny (less than a quarter-inch long) beetle of the weevil family. Like all its weevil kin, it has a long snout. The adult curculio is brown with darker markings and has four prominent humps along its back. The young are white with brown heads, legless, and have slightly curved bodies. They are about three-eighths of an inch long at their biggest. The adults feed on fruit in spring, then lay their eggs in crescent-shaped cuts, which they make in the skin of the apple. After hatching, the larvae tunnel into the fruit.

Plum curculio is a serious apple pest east of the Rockies and is controlled by the application of methoxychlor when petals begin to fall, followed by one or two applications at seven- to ten-day intervals.

Rust Mites

Rust mites are very tiny creatures that are related more closely to spiders than they are to insects. They are brown in winter and white or pale

CURCULIO CATCHER

This strange contraption, using a wheelbarrow and an inverted umbrella, was designed as a weapon against the plum curculio, a type of weevil that feeds on apples. The orchardist smashed the apparatus against the trunk of the tree in hopes of shaking out the bugs, which then would be caught in the umbrella. The author of the book containing this plate did not recommend this method, because bashing the tree hard enough to shake out the bugs was sure to injure the bark of the trunk. From S. E. Todd's *The Apple Culturist*, New York, 1871. (*Courtesy the Library of the New York Botanical Garden*) beige in summer. They become active as soon as new growth appears, causing leaves and fruit to turn a uniform reddish brown.

Control is achieved through the application of the miticide Dicofol or Carbaryl when the mites appear.

Scales

Scales are tiny insects, less than an eighth of an inch in diameter, that have soft bodies and a waxy covering. The immature forms (crawlers) appear in mid-May, move to new feeding sites, moult, then lose their legs, forming the so-called scales. These little insects feed by sucking plant juices and cause discolored spots (often red) on leaves, stems, and fruits.

They can be controlled by the application of Malathion when crawlers are present. Adult scales can be scuffed off the trunk and larger branches of a small tree in winter by taking a piece of burlap or other rough cloth and scuffing the bark as one might use a shoeshine rag. A toothbrush may be used for less accessible spots. This latter method will not destroy all the eggs, however, so it should be followed up with a Malathion spray in the spring.

Spider Mites

Spider mites are also tiny members of the spider clan. Reddish, greenish, or brownish, they usually are found on the undersides of leaves, feeding on plant juices and spinning fine webs while they do so. These little creatures make yellow specks on foliage and leave behind stunted fruit and plants.

To control spider mites, apply Dicofol or Tetrafidon (a miticide) twice, one week to ten days apart, when mites are detected.

Tent Caterpillars

Tent caterpillars are the immature forms of a yellow, yellow brown, or brown moth that has a pair of white stripes on each forewing and a wingspan of about an inch and a quarter. The moths will be seen fluttering about lights at night in June or July. The dark, hairy caterpillar, which can reach two inches in length, has a yellow stripe down its back and often will be found within a web at the crotch of branches. (The fall webworm makes its tents at the ends of branches.) Tent caterpillars feed on leaves and can defoliate a tree.

To control tent caterpillars, apply Malathion or methoxychlor, or remove the webs using the method described for the fall webworm.

4

Apples in Folklore:
The Bright Side

THE FALL
From the
Passio Christi
of Albrecht Dürer,
Nürnberg, 1511.

> Adam lay y-bowndyn, bowndyn in a bond,
> Fowre thowsand wynter thowt he not to long;
> And al was for an appil, an appil that he tok,
> As clerkis fyndyn wretyn in here bok.

Perhaps more than any other fruit, the apple is surrounded by myth and tradition. And the best-known legend is the one that identifies it as the fruit that caused man's downfall in the Garden of Eden.

Just how old this legend might be is hard to say. One notion bandied about is that the biblical story of Eve and the apple (rather than the biblical fruit of the Tree of the Knowledge of Good and Evil) was the creation of poet John Milton in his *Paradise Lost.* But this is pure rubbish. The above four lines quoted in Middle English date from the fifteenth century. Milton was born in the early seventeenth century.

The woodcut by the German artist Albrecht Dürer, pictured here, dating from about 1510, shows Eve taking what is clearly an apple from the serpent's mouth, and the tradition identifying the apple as the fruit of the Tree of Knowledge was old by Dürer's time. In the nave of the great cathedral at Chartres, France, there is a stained-glass window showing Adam and Eve standing with God. In the background is the Tree of Knowledge, the serpent wound around its trunk, its green foliage studded with ruby red apples. The window dates from the first quarter of the thirteenth century.

Despite the antiquity of the tradition, there is not and never has been any physical description in the Bible of the fruit of the Tree of Knowledge that would justify referring to it as an apple. Genesis 3:1–6 (King James Version) reads:

> Now the serpent . . . said unto the woman, Yea, hath God said, Ye shall not eat of every tree of the garden?
>
> And the woman said unto the serpent, We may eat of the fruit of the trees of the garden: But of the fruit of the tree which is in the midst of the garden, God hath said, Ye shall not eat of it, neither shall ye touch it, lest ye die.
>
> And the serpent said unto the woman, Ye shall not surely die: For God doth know that in the day ye eat thereof, then your eyes shall be opened, and ye shall be as gods, knowing good and evil.
>
> And when the woman saw that the tree was good for food, and that it was pleasant to the eyes, and a tree to be desired to make one wise, she took of the fruit thereof, and did eat, and gave also unto her husband with her; and he did eat.

Note: The examples from classical literature in Chapters 4 and 5 were translated by the author from the Greek and Latin texts provided by B. O. Foster in his article "Notes on the Symbolism of the Apple in Classical Antiquity" in *Harvard Studies in Classical Philology*, vol. 10, 1899.

Clearly, there is nothing that says "apple" in that account, and archaeological evidence argues that the apple was unknown in the Middle East at the time the Book of Genesis was written down.

Yet when Christian missionaries told the story of Adam and Eve to the tribesmen of northern Europe, there was no doubt in the minds of those pagan listeners that the Tree of Knowledge bore apples. There was something inherent in their beliefs that pointed to the apple as the culprit.

THE EXPULSION FROM PARADISE
From the *Passio Christi of Albrecht Dürer*, Nürnberg, 1511.

There are many possible explanations for this pagan interpretation. Could it have been that the apple was associated primarily with women? Or perhaps the apple represented a token of love, and the pagans believed that Adam ate of the apple out of affection for his wife? It is also possible that the apple was regarded as a fruit of knowledge, the sacred emblem of some ancient god of wisdom.

To confuse matters, there were two trees in Eden, the Tree of Knowledge and the Tree of Life: "And the Lord God said, Behold, the man is become as one of us, to know good and evil: and now, lest he put forth his hand, and take also of the tree of life, and eat, and live for ever." (Gen. 3:22)

Perhaps the pagans thought of the Tree of Knowledge and the Tree of Life as the same tree and that the apple also was some sort of symbol of immortality.

Finally, we are told, "And Adam called his wife's name Eve: because she was the mother of all living." (Gen. 3:20)

Could it have been that the pagans of Europe once believed in a goddess who was mother of all men and whose sacred emblem was an apple that signified love, knowledge, and immortality?

The answer is a qualified yes, qualified in the sense that we can reach an affirmative answer only through deduction from a lot of circumstantial evidence. Perhaps there never was an apple goddess quite like the one proposed above, but searching for her leads to a yarn as fascinating as any detective thriller.

The Apple and Love

To learn what the apple goddess was like and what aspects of human life she controlled, one might expect to turn to the literature of Druidism, which was the dominant form of worship at the time the Christian missionaries first preached in northern Europe. In this religion, trees were gods, or at least, manifestations of them. Unfortunately, no such literature exists. The tenets of Druidism kept the faithful from writing down their beliefs. Instead, they were preserved in an oral tradition that for centuries was taught by each generation to the next, only to be lost or muddied over in the process of Christianization.

To learn about our apple goddess then, we must turn to other sources, to the remnants of her worship that linger in the folklore of Christian Europe. And we must turn to ancient Greece and Rome, where similar traces can be found.

Very early in history, mankind observed that there were parallels between his life and the lives of the animals and plants around him. Animals were born, grew up, produced offspring, and died. Plants did pretty much the same.

At the same time, perhaps, man also generated the notion that there were spirits behind all these natural activities, and he sometimes imagined these spirits to be personifications of animals and plants and sometimes as genii that resided within the creatures they influenced. This simple form of polytheism was, for Western man, at least, one of the earliest stages of religion, the age of the so-called earth gods who directly controlled the growth of the grain, the fecundity of livestock, and so on.

However, at the dawn of recorded history in the West, with the works of Homer and Hesiod, which were considered sacred histories as much as they were secular literature, we find that the earth gods had already been replaced by the sky gods, a much more intellectual lot, who controlled life on Earth—but only as a sideline, so to speak. In the ancient Greek religion, this was typified by Zeus, the thunder god, who led his brother and sister deities up to Mount Olympus after overthrowing their titan (earth-god) father, Cronos, in a ten-year-long heavenly civil war.

But like politics on Earth, the break was anything but clean, and so we find sky gods with many of the same attributes and powers as the earth gods they had replaced. Apollo, for example, slowly took on the identities of the earlier sun gods, Helios and Hyperion. We find Aphrodite, the goddess of love, being addressed in prayer as the fruitful goddess and the wheat-giving goddess, as if she were one of the earlier vegetation spirits.

Aphrodite's connections with the apple are numerous. When the famous statue of Venus de Milo (Venus is her Roman name) was found early in the last century, a number of fragments were also un-earthed, one of them a hand holding an apple. Pausanias, a second-century historian, describes a statue of the goddess venerated at the Greek city of Sicyon that held a poppy in one hand and an apple in the other. Alcamenes carved an Aphrodite of which a Roman copy exists. It had an apple in its left hand.

Then there are the myths of Atalanta and the Judgment of Paris. The first myth tells how, through Aphrodite's intervention, the fleet-footed maiden Atalanta was beaten in a foot race by Hippomenes.

Briefly, the story goes this way: Atalanta had learned from an oracle that if she married, she would die. However, her father wanted her to take a husband, so Atalanta declared she would marry any man who could beat her in a foot race. She also declared that losers must forfeit their lives, which many did, because she happened to be the fastest thing on two feet.

Hippomenes prayed to Aphrodite for help, and the love goddess provided him with three golden apples. He entered the race, and each time Atalanta left him behind, he threw one of the apples ahead of her, to one or another side of the race course. Each time he did this, Atalanta slowed down to pick up the pretty bauble. She lost the race, of course, and became Hippomenes's bride.

A better-known (and more problematic) apple yarn is the Judgment of Paris, which provides the mythical underpinning of the Trojan War.

According to this story, the goddess Discord was inadvertently left off the guest list for the wedding of Peleus and Thetis, who would later become the parents of the famous Greek warrior Achilles. Discord was miffed, and out of spite, she threw a golden apple amidst the assembled gods and goddesses at the wedding reception. According to the writer Lucian, the apple was thus inscribed: "Fair one, make this your own." Immediately, three goddesses—Hera, Aphrodite, and Athena—laid claim to the token, and after much dispute, the goddesses agreed to let Paris, prince of Troy, act as judge of their respective claims, bringing about the first beauty contest in history.

Each of the goddesses stripped off her robes and displayed her naked charms. And each offered Paris a bribe: Hera promised him

Emblemi d'Amore.

EMBLEMS OF LOVE

This engraving (perhaps nineteenth-century Italian) depicts the ancient Greek legend of the Race of Atalanta. The young princess, Atalanta, stoops to pick up one of the golden apples that Hippomenes, her suitor, has thrown to distract her. Atalanta's desire to have these baubles cost her the race and her maidenhood. The stakes had been high: Atalanta's hand in marriage or Hippomenes's life.

wealth and dominion over Asia; Athena offered military renown and wisdom; and Aphrodite promised him a wife who was the most beautiful woman in the world, namely Helen (later of Troy), who at this time was married to Menelaus, king of Sparta.

Paris chose the love goddess's offer and ran off with Helen. Agamemnon, king of Athens and Menelaus's brother, got together an army and ten years later flattened Troy with the help of the slighted goddesses Hera and Athena.

The story of the campaign against Troy and the city's destruction is found in *The Iliad.* Homer, the author of this great work, does indeed make Helen's running off with Paris the cause of the war. And he does, indeed, have Aphrodite siding with the Trojans, and Hera and Athena siding with the Greeks. However, Homer makes absolutely no mention of either the Apple of Discord or the Judgment of Paris. Surely, Homer wouldn't have overlooked so important a detail as the underlying cause of the conflict.

Actually, Lucian seems to be the first Greek writer to mention the Judgment of Paris and the Apple of Discord, and he lived in the second century A.D. Homer lived sometime between 685 B.C. and 1159 B.C. and apparently knew nothing about the tale. Beyond that, Aphrodite had been depicted in stories and art as giving out apples for at least seven centuries before Lucian's time. Why all of a sudden was she now accepting an apple from Paris?

One possible answer is that later writers made up the story of the Judgment of Paris to explain works of art whose significance had been forgotten. Another explanation (one we will see more of in the next chapter) is that these artworks depicted beliefs of an earlier religion whose tenets were at odds with later religious developments. Perhaps in the second century A.D. there were ancient vases and murals showing three goddesses, one of whom was handing an apple to a young man. No one could quite figure out what the pictures meant, or decided not to say what they meant, and so a new legend was born.

Still, whether Paris gave the token to the love goddess or vice versa, the link between Aphrodite and the apple stands.

Of course, the apple was not the exclusive property of the love goddess. Things are seldom that simple in mythology, because the ancients looked at the things around them from many different points of view, sometimes all at once. For example, the apple tree bore thousands of apples, so the tree could be seen as a symbol of fecundity. Therefore, it would be sacred to a goddess of love and marriage such as Aphrodite. At the same time, the apple tree was something that thrived and bore fruit in summer sunshine, so the apple could be seen as a symbol of the sun's life-giving warmth. As such, it would be sacred to a sun god like Apollo, and indeed it was. Later, we will see how the fruit can be linked to quite a number of different deities, but let us stick with the love goddess for the moment.

The ancient Greeks called breasts "apples," as do the modern Russians (although the latter usage is considered vulgar). Perhaps, then, one reason Aphrodite had a soft spot for the apple might have been that it is shaped like a woman's breast and that both the apple and the breast provide sustenance.

The playwright Aristophanes compares breasts to apples at least three times. In his *Archarnians*, he writes of "breasts that are firm and applelike." In another play, *Ecclesiazusae*, a young girl remarks, in effect, that the flush of maidenhood is on her breasts. Her actual words are very nearly untranslatable: "For something delicate has grown on tender thighs and blooms on the apples." In *Lysistrata*, Aristophanes writes of "Helen's apples," and an unidentified scholar, an ancient himself, has this to say: "They call breasts apples."

Which makes this as good a time as any to remark that a writer we regard as ancient could be writing in the first century B.C. about the so-called Golden Age of Greece in the fifth century B.C., which would be "ancient history" to him. We sometimes think of the ancient world as if it all happened within a year or two. Yet more centuries passed between the building of the great pyramids (circa 2600 B.C.) and Egypt's conquest by Alexander the Great (332 B.C.) than have passed between the birth of Christ and the present. The ancient world changed a great deal during its many centuries. Ancient religions also changed, so that the cherished belief of one generation might have been repudiated a few generations later.

As far as the apple was concerned, things were relatively stable, at least as far as breasts were concerned. Two centuries after Aristophanes, we find the poet Theocritus using the same comparison. In his *Idyll XXIII* we find a young girl repelling the advances of her rustic lover. She cries out: "What are you trying, you little satyr? What will set my breasts afire?" He answers: "I will teach those early apples of yours that they're ripe for nibbling!" (Incidentally, don't expect to find such racy fare in the standard English translations of the works mentioned. They've been cleaned up for domestic use.)

Any culture that could imagine breasts as apples would have little trouble comparing a tree that bears lots of fruit to a woman who bears lots of children. Nor should such a culture have any difficulty in connecting a seed-filled apple with pregnancy. Small wonder then that very early in Greek history the apple began to figure in the rites and customs of marriage.

An ancient historian reports that the famous Greek lawgiver Solon would command a bride to share a Cydonian apple with her bridegroom as a symbol of their union. That was in the seventh century B.C., and his contemporary, the poet Stesichorus, wrote of the Cydonian apples, myrtle leaves, and wreaths of roses showered on a bridegroom's chariot. (For the sake of accuracy, it must be pointed out that the term *Cydonian apple* is usually translated as "quince." However, the ancients did not make the firm distinctions we do among the various members of the apple family, such as the quince, the worden, the medlar, and so on.)

Customs such as eating an apple or throwing apples at a wedding, though different on the surface, probably stemmed from the same basic notion—an attempt to assure the fruitfulness of the newlyweds through the magical use of a fertility token. The rice throwing and use of sugared almonds at twentieth-century weddings spring from similar roots.

From prayers for fruitfulness at weddings, the throwing of apples evolved into a symbol for courtship itself, with lovers said to pitch apples instead of woo, as we might have it. This change apparently happened so fast that only two centuries after Stesichorus, we find the great satirist Aristophanes using the expression flippantly. In *The Clouds*, the allegorical character called Just Reason addresses a young man-about-town: "And don't go running to the theater to gape at everything, for fear some pretty little harlot's apple will hit you and your reputation will be ruined."

Apparently, the custom was long-lived, for two centuries later we find Theocritus mentioning apple tossing, but he presents it in a pastoral setting. In *Idyll V* we find these lines: "And Klearista pelts the goatherd with apples as he drives his flock by, and she whistles at him sweetly." In *Idyll VI* we find a similar description: "Galatea is pelting your little flock with apples, Polyphemus, calling you a goatherd and a man insensitive to love."

Two more centuries go by, and Virgil describes essentially the same scene in his *Eclogue III*, giving Galatea just a touch of the coquette. Polyphemus says, "Galatea hits me with an apple, then hides among the willows, but she really wants me to see her."

In the hands of the crafty Romans, the apple went swiftly from a ballistic missile to something a lot more tender. Propertius composed the pretty little lyric that follows, addressing it to a girl he called Cynthia. (Her real name was Hostia, and she and the poet were lovers from 28 to 23 B.C.) The poet has discovered Cynthia asleep:

And I loosened the crowns from our foreheads,
And I put them, Cynthia, to your temples;
And I reveled in loosening your hair;
Then I put an apple (I had secreted them in) into the hollow of each of
    your hands,
And I lavished all these things on thankless sleep,
Gifts that often roll from an upturned pocket!

Catullus, a few decades before, also wrote of a gift apple that rolled out at the wrong moment. The poet addressed these lines, translated here in prose, to his friend Hortalus:

Hortalus, I send you these translated verses of Callimachus [a Greek poet], lest you think . . . I have forgotten your words totally, like the apple sent secretly by a lover that rolls out of the bodice of a chaste maiden. The poor child had concealed the fruit beneath her gown, then had jumped up at her mother's arrival and shaken it out of her clothing. The rushing thing, rolling down, colored a sad face a very revealing red.

Finally, we might want to close this part of the discussion of ancient apple mythology with an epigram attributed to Petronius, the most distinguished courtier of the emperor Nero: "Send kisses with apples, and I will feast with pleasure!"

So if we start counting with Solon in the seventh century B.C., the link between love and the apple had already survived at least 700 years when Nero came to power. But there is no reason to think the idea was new to Solon, and we know it survived for many centuries after Petronius.

A very recent example can be found in Newbell N. Puckett's *Folk Beliefs of the Southern Negro.* There it is stated that early in this century apples were among the magic objects used in New Orleans voodoo ceremonies when love was the goal. That makes 2,600 years that love and the apple were tied. Old beliefs die hard, if they die at all.

So far in our search for the apple goddess we have established two major points: The pagan mind associated the apple with women and with love; and specifically, the apple was used as a symbol of a woman's breasts, and it was a token of the love goddess. Because of the love-goddess association, the apple figured in ancient marriage ceremonies and courtship customs, and it became a love gift or token in its own right.

Now we can turn to what must be the apple's most widespread and persistent love connection. That is its use in answering lovers' questions. However, a brief side trip first.

We are probably all familiar with the old daisy ritual of "She loves me; she loves me not. . . ." But why should the daisy have the answer? Well, it seems that long ago people believed that the daisy had a special relationship with the sun. The name of the flower was originally Day's Eye, for during the day, the flower's white petals stay open, revealing the yellow "eye." At night, they close and cover it. It happens that the sun looks down on the world from the sky, and people once believed that it "knew" everything, so any flower that "watched" the sun all day just had to pick up some of that knowledge. Lovers couldn't very well address their questions to the sun, but they could ask the daisy.

It was the same sort of thing with the apple. The love goddess was hard to talk to, but her symbol, the apple, was very accessible. It was the love goddess who decreed who should marry whom, and the apple could be used to divine her will.

The Latin author Horace mentions apple divination in his *Satires*, written about 35 B.C. In one of them he asks young lovers, "How can you be in possession of your minds, you who pick the seeds from apples, then delight in bouncing them off the ceiling?"

A later Latin commentator explained Horace's barb: "Lovers customarily squeeze apple seeds between their first two fingers, shooting them at the ceiling and thus auguring whether they can hope to realize their dreams."

In 1920, essentially the same ritual was reported in Daniel L. Thomas's *Kentucky Superstitions* as a folk custom in that state. Young people there would give each seed the name of some boy or girl. Then they would shoot the seeds and determine the names of their future mates according to which seeds reached the ceiling.

The Kentuckians used several variants of this custom as well. A lover could give names to several apple seeds, then stick them to either his or her forehead or eyelids. The first or last seed to fall, according to local custom, would be the future spouse. Or the named seeds could be placed on a hot stove, and the first to jump or the last to burst open would be recognized as the true and steadfast lover.

There also was a counting ritual that was something like plucking a daisy. One would eat an apple, save all the seeds, and count them out according to this formula:

> One, I love;
> Two, I love;
> Three, I love, I say;
> Four, I love with all my heart;
> Five, I cast away;
> Six, he loves;
> Seven, she loves;
> Eight, both love;
> Nine, he comes;
> Ten, he tarries;
> Eleven, he courts;
> Twelve, they marry.

Naturally, the apple's powers of divination in the affairs of love were not restricted to the seeds, since the entire fruit and even the apple tree itself were sacred to the love goddess. An old German folk tradition held that if an apple tree bloomed in autumn, it was the sign of an approaching marriage. Further, the number of divination rites using the entire apple or its peel, if anything, is greater than for those using the seeds alone.

The business of ducking for apples on Halloween did not begin as the children's game it is today. This apple ceremony was part of another celebration, and the intent was quite serious. Each year in Ireland, the Celtic natives used to kindle a new fire on Halloween, which they called the Eve of Samhein. The First of November, called Samhein, was the Irish New Year. The new fire, it is said, was used to light afresh every hearth in the country. The hope was that the influence of this new and blessed fire would last through the entire year.

On the Eve of Samhein, the Irish would try to divine their destinies, especially their fortunes during the coming twelve months. For the young people this usually boiled down to whether and whom they would marry. The apple, of course, held the answer.

An eighteenth-century writer, Charles Vallancey, described Irish fortune-telling this way:

> They dip for apples in a tub of water, and endeavour to bring one up in the mouth: they suspend a cord with a cross-stick, with apples at one point, and candles lighted at the other and endeavour to catch the apple, while it is in circular motion. These and many other superstitious ceremonies, the remains of Druidism, are observed on this holiday, which will never be eradicated while the name of Saman [Samhein] is permitted to remain.

Youngsters in twentieth-century Maryland seemed to have named the apples just as the Irish once did. This description is found in A. W. Whitney and C. C. Bullock's *Folklore from Maryland:*

> On Hallowe'en, put some apples in a tub of water, and name them with a label. Let a girl kneel over the tub, shut her eyes, put her hands behind her, and try to catch the apple with her teeth. The one she succeeds in catching will be her future husband.

Sometimes the apples were left unnamed and the question asked then was, "Will I marry? Yes or no?" The answer depended on whether or not one caught an apple. Vallancey's apple and candle routine, mentioned above, was probably a yes-or-no ritual. The cross-stick he mentions was probably suspended from a string. Two of the arms of the cross-stick bore apples while the other two held lighted candles. Someone would twist the string and whirl the cross, and someone else, with eyes closed, would try to catch something with his teeth. A mouthful of greasy candle or a burnt nose wasn't nearly so hopeful a sign as a bite of sweet apple.

While bobbing for apples was suitable for parties, the ritual that follows was carried out under spookier conditions—the questioner was all alone in a darkened room at midnight on Halloween. The description is from James George Frazer's *The Golden Bough*, and the belief is old Scots:

> You took an apple and stood with it in front of a looking glass. Then you sliced the apple, stuck each slice on the point of a knife, and held it over your left shoulder, while you looked in the glass and combed your hair. The spectre of your future husband would then appear in the mirror stretching forth his hand to take the slices of apple over your shoulder.

Finally, let's take a look at what had to be the most elaborate way of divining with an apple, or more specifically, with its peel. This custom was known to the Irish and Scots and exists in many areas of America. This particularly elaborate version of the ritual can be found in *Folklore from Maryland*.

> On Halloween, cut the rind of an apple, throw it over your shoulder, and it will form the initial of your lover's name. The correct way to do this is to select first a large apple:
> It must be carefully pared and swung cautiously around the head. If the peel breaks, there is calamity ahead and an interrupted love. If it breaks twice—once while being pared and again while being thrown—then it means still more disaster; for a twice-parted peel means twice-parted love. The peel must be twirled around the head three times and then must be tossed lightly from the finger. It will find its way in a great swirl to the floor, and there it will lie, ready to be read. If the moon

is full, the peeling must be read from the north; if the time be midnight, it must be read from the south; if the night be rainy, it must be read from the east, and if it be clear, it is read from the west. The letter is studied out, and the girl knows what initial will be embroidered on her linen at her marriage.

The Apple and the Sun

In a sense, we have established one more point in the search for our hypothetical apple goddess. If the apple was sacred to the love goddess and could be used to divine a lover's fortune, then the fruit possessed a knowledge of sorts and therefore was a pagan "fruit of knowledge." However, this quality of the apple could have come from another source: its sacredness to the sun god Apollo. As mentioned in connection with the daisy, the sun or sun god was thought to know everything because he could see everything from his heavenly vantage point. Perhaps this second association played a part in making the apple a "fruit of knowledge."

Unhappily, there are fewer specific links between Apollo and the apple than there are between the apple and Aphrodite. We know that the apple tree was one of several trees sacred to the sun god. Others were the laurel, the tamarisk, and the poplar, this last inherited from the older sun god Helios, whose worship Apollo had supplanted. We also know that the sanctuary of his shrine at Delphi was decorated with garlands of fruit bound to laurel boughs. However, there is archaeological evidence that Apollo seized the shrine at Delphi from the moon goddess Artemis and that the apple was her symbol first.

Beyond that, there is the suggestion that the name *Apollo* comes from the same root that gives us the modern English word *apple*. Actually, Apollo as we know him was probably a composite of several ancient gods. By classical times, he had pretty much absorbed the old sun god Helios, and he was also identified as the god of the Hyperboreans, "the men who live beyond the North Wind," whom the ancients believed were the inhabitants of a happy, sunny clime where sickness, old age, and sorrow were unknown.

Although we know better, many of the ancients believed that

Britain was this sunny Never-Never Land, and it just happens that the word for apple in ancient Britain was *abul*, which is quite close to *Apollo*. What is particularly interesting is that the Celtic tribes of Britain also believed in this happy kingdom of the sun. They called it the Isle of Apples, or Avalon, and it was here that King Arthur supposedly went to spend eternity.

The links between the apple and Apollo are admittedly weak, but it is clear that some existed. Furthermore, while Apollo is the most familiar of the pagan solar deities, it is the tie between the apple and the sun that really interests us. There was a solar divinity sacred to the ancient Latvians and Lithuanians who actually was described as an apple on at least two occasions. The few traces that remain of this ancient figure are found in a series of charming folk songs called *daina*, which preserve a group of solar myths.

Among the Letts and Lithuanians, the sun was feminine and the moon masculine, just the reverse of the tradition that has come down to us from Greece and Rome. And the ties between the Baltic sun goddess and the apple are legion. In one *daina*, we find the rays of the sun represented as an apple tree with nine branches. In another song, we find the sun-goddess weeping because the golden apple has fallen from the tree, a poetic evocation of sunset. Finally, in one *daina*, the sun goddess is called a silver apple, and in another she is an apple sleeping in an apple orchard, bedecked with apple blossoms—the pinkish clouds of sunrise. The apple here is not merely the token of the sun; it is the sun herself.

Equally interesting, there also is in the *daina* an explanation for this equation of sun with apple. Both day and night are envisioned as trees, and the various celestial bodies are the fruits of these trees.

Just as interesting is the fact that a thirteenth-century Welsh poet, Dafydd ap Gwilym, uses the same image. He calls the stars cherries on the tree of night, the plums of the unloved moon, fruits grown in her frosty orchard. The idea was by no means unique to the Baltic.

Turning to Celtic legends for a moment, we can find two instances in which the apple is associated with solar figures—ancient heroes rather than gods per se. In "The Cattle Raid of Culange," the Irish hero Cuchulainne, a solar figure, has set out to learn the art of war. On his journey he must cross the Plain of Ill Luck, a place where men's feet first stick to the ground, then are caught by the blades of grass. (Cuchulainne's body gave off so much heat that it melted the snow for thirty feet around. Once, when he was enraged, he was thrown into three caldrons of cold water, one after the other. The first caldron simply burst from the heat; the second turned to steam, and the third was heated lukewarm before Cuchulainne had simmered down to a point where he could be reasonable.) Now when Cuchulainne went to cross the Plain of Ill Luck, he got instructions from a youth who told him to follow the track of a wheel, then that of an apple that would roll before him. The wheel, of course, symbolized the disk of the sun. The apple also was a sun symbol. In short, the only safe path for a sun hero was the path of the sun itself, and the apple would "know" what that was.

In the story of Kulhwch and Olwen, found in a collection of Welsh tales called *The Mabinogion*, there is a similar link between sun, wheel, and apple. Kulhwch has gone off to woo Olwen, whose name comes from the Welsh word *olwyn*, meaning "wheel." Now we have said that the wheel was a sun symbol, but this isn't the only clue to Olwen's real identity. First, she had a rosy complexion and her beauty was unsurpassed. Secondly, her dress was flame red, and around her neck she wore a collar of red gold. Gold itself was a symbol of the sun, gold rings and collars even more so.

The apple is found in Kulhwch's costume. The description comes from Lady Charlotte Guest's 1848 translation: "About him was a four-cornered cloth of purple, and an apple of gold was at each corner, and every one of the apples was of the value of an hundred kine [cows]." Could there be a more suitable emblem for a man who would woo the wheel of the sun?

These Celtic legends are many centuries old, but folk customs from more recent times hint at the same sun-apple links.

Until the mid-nineteenth century, country folk in Belgium observed the following custom: Every year on the first Sunday of Lent, the young people would take lighted torches and run with them through the orchards, shouting:

> Bear apples, bear pears,
> And cherries all black
> To Scouvion!

Then they would whirl the torches and hurl them among branches of the trees. It seems that the natives of the district called the first Sunday of Lent the Day of Little Scouvion. The next Sunday was dubbed the Day of Great Scouvion, and the same ritual was repeated.

The explanation of this custom would seem to be that Scouvion (or something like it) was the name of some ancient solar deity worshiped in that region. The torches were either reminders to the trees of the approach of summer, symbolized by the flames, or a demonstration to the god Scouvion of what his worshipers expected of him, namely that he shed his warmth on the trees and make them bear fruit.

Finally, there is the custom of the wassail bowl, the traditional British drink of the holiday season. The word *wassail* comes from two Middle English words, *waes* and *haeil*, meaning literally "Be hale!" In other words, "Be hale and hearty!" It was a wish for good health, a toast one man made to another when he drank from the wassail bowl.

Wassail usually was drunk on Christmas Eve and Twelfth Night, at the beginning and end of the twelve-day Christmas festival of centuries ago, the "twelve days of Christmas." The liquor was a mixture of hot ale and spices, sometimes laced with wine or brandy. Traditionally, roasted apples were floated in the brew.

It is hard to believe that the custom of drinking wassail, though Christianized as part of Christmas festivities, wasn't a remnant of some earlier pagan custom. Christmas falls at the time of the winter solstice, the shortest day and longest night of the year. The sun would be seen as being at its weakest moment.

What with a round bowl, hot ale, and roasted apples, it is hard to believe that our ancestors weren't drinking with the sun in mind, that they weren't drinking to the sun's good health as well. The apple in the wassail bowl carried a special significance. It was a symbol of the sun god Apollo, whom the Greeks regarded as a healing god. It was a token of the Isle of Avalon, where battling heroes went to be healed of their wounds. The apple itself carried a symbolic wish for good health.

To your health, old Sun! Wassail!

5

Apples in Folklore:
The Dark Side

This late medieval woodcut of the Judgment of Paris clothes participants anachronistically but otherwise conforms to the ancient Greek legend. The three goddesses (Aphrodite, Athena, and Hera), dressed as noblewomen, and the god Hermes, holding the messenger's caduceus but with a crown replacing his broad-brimmed hat, come upon the reclining Paris, shown here as a knight in armor. From Giovanni Battista Cantalicio's *Judicium Paridis*, Wittenberg, 1514.

Up to now, our search for the apple goddess has led us to the temples of her "brighter" descendants, the love goddess and the sun god. The gods we will visit now are her darker, more mysterious children.

But before we can proceed, we must consider one point. The history of ancient man covers thousands of years, and during those years, the organization of human society changed profoundly, and so did the myths that ancient man used to explain why things were the way they were. The blessed gods of one millennium could become the devils of the next. The heroes of one generation could become the dullards of another.

The gods as they are commonly known to us today reflect rather late stages in the development of the pagan religions of Greece and Rome. The love goddess was once a far more powerful divinity than the discussion of Aphrodite in the last chapter would indicate. By the time the first myths were written down, she had been reduced to the American Dream, beautiful, nubile, and having the morals of an alley cat. No questions asked, she bedded down with any god or man that appealed to her.

Yet despite the efforts of the later mythmakers, there was something sinister lurking behind her beautiful face. The Athenians identified her as the eldest of the Fates, the three goddesses who controlled men's lives. In other places she was called the black goddess, the man-killer, and Aphrodite of the Tombs. These ideas about the love goddess are not just droll allusions to the kinds of heartaches men and women can have in love affairs. They hint at the far more powerful position that the love goddess once enjoyed.

The same thing happened to our apple goddess. After she and her kind fell from power, there was an attempt, doubtlessly deliberate, to wipe out any trace that she had ever existed. Her powers over the lives of men were parceled out to other gods and goddesses. The stories in which she had once been the moving force were altered to conform with later religious ideas in which she had no part. But the stories themselves did survive, no matter how altered, and through them we can find the apple goddess.

The search will not be easy, however, because at each step of the way, we will have to dig through layers of later myth to find the earlier meanings beneath. First, we will examine the beliefs surrounding the moon goddess Diana, because she will take us one step closer to our goal. Then we will visit with the vegetation god Dionysus, and finally, we will examine the worship of the hero-god Hercules, who will bring us to the sacred grove of the goddess herself. There we will find the explanation for the apple's persistent hold on the imagination of Western man.

Diana, the Moon Goddess

In the heyday of their culture, the Romans had a moon goddess named Diana, whom they thought of as the twin sister of their sun god, Apollo. Diana had a festival in mid-August, and that day the Romans served a ritual meal consisting of roast kid, cakes served on plates of leaves, and apples still hanging on their boughs. We know that the apple was sacred to her brother, Apollo, so it should not be altogether surprising that it appears in her worship as well. When two gods are considered to be tied by blood or marriage, they often share the same symbols, which is another way in myth of saying that two gods are closely related. For example, both Apollo and Diana were renowned as archers.

Unhappily, the ancient Romans did not say exactly why the apple was sacred to Diana, but there is no doubt that she was a moon goddess and that the moon was felt to play an integral role in the lives of growing things on earth. Nor did this kind of idea die with the ancients.

The 1974 edition of *The Old Farmer's Almanac* devotes an entire page to a planting timetable based on the phases of the moon. The almanac suggests planting flowers and vegetables that bear their crops above ground during the "light" of the moon—from the day of the new moon to the day of the full moon. Plants that bear their crops underground should be planted during the "dark" of the moon—from the day after the full moon to the day before the new moon.

Other twentieth-century folk beliefs expand well beyond the sphere of planting. In Maryland, farmers believed that if apples were picked during the dark of the moon, they would keep better than if they were picked during the light. Nor would these farmers prune their fruit trees during the dark of the moon for fear the trees would turn black and rot.

Beyond that, a relationship between the moon and growing things seems to permeate folklore around the world. Very early in human history man perceived that there was some tie between the moon and the growing season of plants. Suppose that one year the tree buds began to open during the days of the full moon. Odds are that thirteen full moons (364 days) later the tree buds began to open again and that thirteen full moons after that the same thing happened once more. Soon ancient man began to believe that it was the moon that opened the tree buds.

Later on, man realized that the yearly growth of vegetation was related to the sun; that during the summer the sun seemed to rise higher in the sky and to stay there longer, shedding more warmth on the earth below and making the plants grow. But in the beginning, it was the moon that counted, and the first calendars that man devised were based on the cycles of the moon.

But enough lunar lore for the moment. The god Dionysus is waiting, and we will return to the moon once more, but not until we have met the "dying-and-reviving" gods, of which he is a prime example.

Dionysus, the God Who Dies

Dionysus is most familiar as the god of the grapevine, but he was also a tree god. The historian Plutarch writes that most Greeks offered sacrifices to him as "Dionysus of the tree," and in Greek art he was often represented as an armless, upright post, draped with a mantle and fitted with a bearded mask for a head. The cut ends of leafy boughs were tucked behind the mask, giving the figure a treelike appearance.

Dionysus was the patron of all cultivated trees; prayers were offered to him to make trees grow, and fruit growers set images of him in their orchards in the form of natural tree stumps.

A specific reference to the apple can be found in the writings of the third-century writer Athenaeus, who credited Dionysus with the discovery of the apple, while many centuries before, in the poem "The Pharmaceutriae," Theocritus wrote of a young lover who tells his lady friend that he should have come to her "bearing the apples of Dionysus in the folds of his robes." A later Greek commentary on Theocritus's poem says that the apples Aphrodite gave Hippomenes in his race against Atalanta were apples taken from the wreath that Dionysus wore around his head.

Dionysus really represented all vegetation that seems to die in autumn only to grow anew in spring. The myth of Dionysus represents this process of death and revitalization as the product of a power struggle among the gods. The story usually was told this way: Dionysus was fathered on a mortal woman by Zeus, the father of the gods. Zeus's wife was jealous and had her henchman tear the infant Dionysus to pieces. Afterward, the pieces of the little god were put back together, and he was restored to life.

Anyone who has seen what a vineyard worker does each autumn to a grapevine, cutting off nearly every limb, understands the myth. And the grapevine, after all, was the god's primary emblem. Actually, one of the most graphic connections between the apple and a dying-and-reviving god like Dionysus was preserved in a folk custom of nineteenth-century France. The description that follows comes from Frazer's *The Golden Bough.*

> In Beauce, in the district of Orleans, on the Twenty-fourth or Twenty-fifth of April, they make a straw-man called "the great *Mondard*." For they say that the old *Mondard* is now dead and it is necessary to make a new one. The straw-man is carried in solemn procession up and down the village and at last is placed on the oldest apple-tree. There he remains till the apples are gathered, when he is taken down and thrown into the water, or burned and his ashes cast into water.
>
> Here the straw-man . . . represents the spirit of the tree, who dead in winter, revives when apple blossoms appear on the boughs. Thus the first person who plucks the first fruit from the tree and receives the name of "the great *Mondard*" must be regarded as a representative of the tree-spirit.

The last sentence of that passage is pregnant with meaning, because it hints at a very ancient custom. The first person who plucks the first fruit "receives the name" of the tree spirit. It is a reference to an ancient notion of alternating kingship that will appear again, with darker significance, as we get nearer to the worship of the apple goddess.

A second Greek god, Adonis, also died and was revived. According to myth, he spent each summer on earth and each winter in the underworld. What is particularly interesting about this god, however, are the somewhat contradictory details of his parentage. Adonis was a foreign god adopted into the Greek religion and, as often is the case in such adoptions, he was made a child of the reigning deity of the local pantheon, in this instance, Zeus. However, certain elements survive in the legend of Adonis that may hint at his true origins.

Adonis had a foster father, Melus, who hanged himself out of grief when he learned of Adonis's death. In life, Melus had been a priest of Aphrodite and, in death, he was transformed into an apple tree by the love goddess out of pity. It just happens that *Melus* (spelled *Melos* in Greek) is almost identical with the Greek word for apple, *melon.* And there is a chance that the story of Aphrodite's turning Melos into an apple tree was a myth made up to explain why fruits played a part in ceremonies marking Adonis's death. Theocritus describes one of these rituals, writing: "Before him [Adonis] lie all the ripe fruits that are borne by trees and in delicate gardens, arrayed in silver baskets."

However, the ancient author who recounted the tale of Melus's transformation may have been thinking of another myth, now lost, that had Adonis's mother, Smyrna, bearing the god after eating or otherwise coming in contact with an apple. Perhaps Melus (the apple) was the father of Adonis in some original version of the myth, and not his foster father. This suggestion is not so farfetched as it might seem. Consider the case of Attis, a god who closely resembled Adonis and, like him, came from Asia Minor. Attis was conceived after his mother, a virgin, placed a pomegranate in her bosom. No doubt as a result of that belief, it was taboo for the devotees of Attis to eat pomegranates, apples, or dates.

The notion of impregnation by a piece of fruit was hardly limited to the ancient world. A folk tale originating in Venice, called "Apple and Apple Skin," is based on the same idea. In this tale, a nobleman and his wife are childless, and one day the nobleman meets a sorcerer and asks what must be done if his wife is to conceive. The sorcerer gives the husband an apple and tells him that his wife will conceive nine months after eating the fruit. The nobleman gives the apple to his wife, who eats it after first asking her lady-in-waiting to peel the fruit. The lady's attendant eats the peel, without her mistress knowing it, and nine months later, each is delivered a son. The mistress' son is as white as the flesh of an apple, the maid's son as ruddy and white as the peel.

We cannot abandon this masculine role of the apple without mentioning the fact that in England the apple tree replaced the maypole, that ancient phallic symbol of Europe, after it had been banned by the Puritans in the seventeenth century. In spring, the young people would sprinkle an apple tree with cider, then circle it in dance, singing their hopes for plenty in the coming harvest.

The form of that ritual, at least, is clearly related to another custom where farmers in certain parts of Britain would "wassail" their apple trees. The custom survived well into this century and may still be around today. Here is how A. J. Downing described it in 1845 in *The Fruits and Fruit Trees of America*:

> As the mistletoe grew chiefly on the apple and the oak, the former tree was looked upon with great respect and reverence by the ancient Druids of Britain, and even to this day, in some parts of England, the antique custom of saluting the apple trees in the orchards, in the hope of obtaining a good crop the next year, still lingers among the farmers of portions of Devonshire and Herefordshire. This old ceremony consists of saluting the tree with the portions of a wassail bowl of cider, with a [piece of] toast in it, by pouring a little bit of cider about the roots, and even hanging a bit of the toast on the branches of the most barren, the farmer and his men dancing in a circle round the tree, and singing rude songs like the following:
>
> Here's to thee, old apple tree,
> Whence thou mayst bud, and whence thou mayst blow;
> And whence thou mayst bear apples enow,
> Hats full! Caps full—
> Bushels and sacksfull!
> Huzza!

This last custom may be related to Dionysus. But as we saw in the last chapter, wassail could be linked to such sun gods as Apollo. Well, there is another possibility. Perhaps the custom of "wassailing" apple trees grew out of the worship of a god who combined aspects of both Apollo and Dionysus. The Greeks and Romans had such a god, Hercules, who began as a mortal, led a life filled with heroic exploits, died, and emerged as a sun god after death.

Hercules, the Man-God

In one of his exploits, Hercules had set out to steal three of the apples of the Hesperides. These were the golden apples that Mother Earth gave to Hera, queen of heaven, on her marriage to Zeus. Afterward, the apples had been entrusted to the care of three minor goddesses known collectively as the Hesperides. Here, the myth gets a bit complicated.

The Hesperides were the daughters of Atlas, the god who held the world on his back, and a goddess we really should call Evening, because that was what her name (Hesperis) meant. Hercules was stumped over how to get the apples, so he asked Atlas to give it a try, agreeing to hold up the world while Atlas attempted to wheedle the fruit from his daughters. Atlas succeeded and would have left Hercules holding the world forever, had the hero not tricked the god into taking back the burden.

The number of hidden references to the sun (and, subsequently, apples) in this little story are many. First off, apples themselves can be solar symbols, as we saw in the myths of the Latvians and Lithuanians. Second is the notion that they were the *apples* of the Hesperides. In the myth, the Hesperides are the children of Atlas and Evening, but the ancients also imagined the Hesperides as islands that lay just beyond the western horizon, at the point where the setting sun sank into the sea.

Finally, there is Atlas himself, who very likely was a sun god before Zeus became king of the gods. The myth of Atlas's supporting the world on his back may have been a later reinterpretation of art works that show him holding a solar disk. The fact that he was married to Evening also argues for a sun god identity. One thing is sure, however. Atlas clearly was a powerful deity in the religion that preceded that of classical Greece. In the celestial civil war mentioned in the last chapter, Atlas was the leader of the earth gods. This argues that in the older, pre-Olympian religion Atlas held the same rank as Zeus, whose name can best be translated as "bright sky."

The point is this: Zeus can be regarded as a sun god who replaced another sun god (Atlas), and if Hercules could replace Atlas as bearer of the world (meaning sun disk), even if for only a few hours, then Hercules can be regarded as having the power of a sun god or as being a sun god in his own right. In all likelihood, Hercules was the sun god of some city on the Greek peninsula centuries before the arrival of the Greek-speaking tribes. (The same would apply to Atlas.)

The idea that a mortal king such as Hercules could become a sun god after death probably was based in the religious beliefs of the New Stone Age people who lived around the Mediterranean and across most of Europe before the first Greek tribes reached Greece. That was about four thousand years ago.

At the end of the New Stone Age and down into the Bronze Age a form of religion developed in which the moon reigned supreme. And there was good reason for putting the moon ahead of the sun, who took priority in the later Iron Age religions. The moon, shining against the dark night sky, was more easily recognizable as an entity, though at first man seems to have thought of the moon in its different phases— new moon, full moon, old moon, etc.—as separate moons.

Still, it was the moon that formed the basis of the calendar. It was the moon that seemed to regulate the seasons and all seasonal events, such as the growth of crops. For a variety of reasons, the twenty-eight-day menstrual cycle of the female among them, the moon was personified as a goddess.

The rising and setting suns doubtlessly were noted quite early, but since they bracketed the darkness, they were likely to be associated with night and the moon. The daytime sun is lost in the general brightness of the sky, and perhaps for this reason man seems to have recognized it as a distinct celestial body much later in history. When the sun finally was recognized, it was cast as second lead in the yearly drama, as husband to the moon goddess.

Another feature of this very ancient religion was the moon priestess, who apparently was regarded as the daughter of the moon goddess and as the actual incarnation of the moon goddess on earth. Further, like her mother, the moon priestess was married, and her husband was

*Del pigliare dell'Arbore le palmuccie, quando
l'Arbore ha i frutti, & conseruar-
le buone da inneftare.
Cap. XXXVI.*

Apple picking, seventeenth-century style. From Marco Bussato's *Giardino D'Agricultura*, Venice, 1612. (*Courtesy the Library of the New York Botanical Garden*)

regarded as the son of the sun god and as the incarnation of the sun god on earth.

When the ancients finally observed that there was a sun, they very quickly noticed that the sun went through two long cycles each year. From the winter solstice to the summer solstice the days got longer. Each day the sun rose higher into the sky and seemed to stay there longer. However, from the summer solstice to the winter solstice the process was reversed.

To explain this, these people believed that there were two suns: a waxing sun, who reigned from the winter solstice to the summer solstice, and a waning sun, who reigned the other half of the year. Here is where the unpleasantness began. These people also believed that the waxing sun was annually displaced by means of a bloody coup, that he was murdered by his successor. The same, of course, applied at the other half of the year, and therefore the same had to apply on earth.

Every six months, at the solstices, the bloody coups of heaven were enacted on earth. The sun king, the incarnation of the waxing sun, was ritually slain by his successor, the incarnation of the waning sun, who was not regarded as a king. The successor, in turn, was slaughtered by the succeeding sun king, and so on.

To compensate the sun king for the rude treatment given him by his successor, the ancients allowed the sun king the privilege of immortality. For some reason, they did not extend the same courtesy to his successor.

Finally, they came upon the idea that it was the same sun that existed during the entire year, and then that it was the same sun that shone every year. This cut down on the need to sacrifice the sun king each year, which was just as well, because whereas the sun might die and spring back, the sun god's earthly incarnations died quite a bit more readily than they revived.

In any case, after the sun king died, he was figuratively reborn in the person of his successor, and he was spiritually reborn by his soul's joining the spirit of his father, the sun god. In short, the yearly ritual became the doorway through which a mortal sun king could reach immortality and godhood. This brings us back to Hercules, who surely began as a sun king of this sort, and to the apple.

First, we know that Hercules was entitled after death to burnt sacrifices in his honor, and the usual victim in sacrifices to him was a bull, a symbol of strength and male fertility. However, when no bull was available, there was an alternative victim—an apple tricked out with four stick legs and a pair of stick horns. It could be argued that this was nothing more than a substitute bull, but then, why make it out of an apple? The answer seems to be that the apple played an integral part in the process that converted the mortal Hercules into a god. The apple's part may have been changed when the myth of Hercules was reconstituted by the Bronze Age Greeks, but it was, after all, changed, not eliminated entirely.

Remember the apples of the Hesperides, the three golden apples that Hercules was going to steal? Well, he wasn't going to steal them originally. The theft idea is without doubt a later Greek myth grafted onto a New Stone Age or early Bronze Age original. The original version would have gone something like this: Hercules, the sun king, was at the end of his reign. He went to the sacred grove of the moon priestess, and there he was given a ritual meal of apples, the last meal of a condemned man. The apples were a solace to him, however, because by eating them he was blessed with the knowledge that life and death were simply two sides of the same coin. In eating the apples he was assenting to his fate and gaining immortality. Then he was ritually tortured, slain, and his body was burnt on a pyre. And Hercules was in this way united spiritually with his father, the sun god, because in life Hercules had been the son and incarnation of the sun. The mortal man had become a god.

This may sound farfetched, but there is evidence that the apple played an integral role in the process that turned a man into a god.

The late New Stone Age and Bronze Age people apparently believed that similar ritual murders including apples as a final feast would ensure the growth of crops, which roughly follow the same sort of death-and-revival cycle, though these ancients didn't know why. In short, the death of the sun king was imagined as a benefit to the entire community. But the question remains why the apple should play a pivotal role in this kind of belief.

The answer to that question is conjectural, since we're trying to imagine what an entire tapestry was like when all that remains are a few threads. To complicate matters, these threads were reused to spin new yarns, to weave new pictures.

A Myth of Creation, Perhaps

Perhaps the apple was the chief or favorite form of sustenance of some New Stone Age tribe. Perhaps this tribe lived where apple trees were common and the people identified themselves with the apple, considering themselves its children.

Surely these ancients would have noticed the moon, and perhaps they imagined the moon as a silver apple. They also would have noticed the setting and rising suns, and perhaps they imagined these suns as golden apples. Or maybe they imagined their suns as the golden cheeks of a silver moon-apple, like the fruit that stays pale and waxy in the shade but offers a rosy cheek to the light.

Even after these people had noticed the daytime sun and realized that it was also playing a part in the yearly drama of the seasons, they would have stuck by their moon, their silver apple, so the sun could only share in the moon's powers, at most. The moon, personified as a goddess, would be their chief provider. She controlled the fertility of the seasons and their own human fertility. She would be their mother, their apple goddess.

Just how these people might have envisioned the relationships among their gods or the origins of the world is a matter of pure speculation. They would have had a creation myth, and one is tempted to reconstruct the tale something like this:

In the beginning, there was Earth and Sky.

And Earth produced the tree of night, which grew up to Sky, and the tree bore an apple that was Earth's daughter. And the daughter was called Moon, and she was given dominion over the lower branches of the tree, which were the Earth.

And Moon divided Earth into different regions, making some land

and some water, because before this, Earth had been shapeless. And she divided Sky into night and day, because this too had been shapeless.

And she showed her bright face by night and her dark face by day, and she showed her red face between the day and the night, because she was queen of all the sky.

And Moon showered her silver dew on the land and filled the land with green things. And she created the apple, which was like her own face, light and dark and red. She created the apple, which was like her own breast, round and swelling.

Then Moon became lonely because there weren't any living things. She saw her silver face on the dark back of the sea and divided her reflection into a thousand silver fishes. And she blew upon the land, and Earth produced a thousand crawling things, and Moon's breath filled the air with a thousand flying things. But still she was lonely.

So Moon went to Mother Earth and asked for a companion. So Earth produced the tree of day, which bore Sun as an apple, and his face shone like Moon's.

And Moon gave Sun dominion over day, so he might bring light forever to all the green things and living things on the land. But he was pale, like a man, and he began to die.

So Moon gave Sun one of her apples, which were her breasts, and he took sustenance thereof and he was reborn. And Moon gave Sun her golden cheeks, and now Sun was like Moon, because he had tasted of her apples. And they lived a thousand years together, and this was before there were any men.

And Moon and Sun had many children, which were the stars, and together they were the parents of all men, whom they created male and female in their own image.

And the gods put the men in the garden Moon had planted, and they gave the men all manner of good things to eat. And in the center of the garden grew the apple tree, and when the men ate the apple, which was light and dark, they knew that life and death were two faces of their mother, Moon. And when the men ate the apple, they knew they were immortal, for like their father, Sun, they could be reborn. . . .

Obviously, this creation myth was consciously contrived in an attempt to prove the existence of the apple goddess. However, a good piece of fiction is based on facts, and this one is. We know, for example, that the apple was sacred in the worship of both the moon goddess Diana and the sun god Apollo. Further, we have reason to believe that the apple was moon's token first, and that she gave it to the sun.

The idea that the apple belonged first to the moon comes from several different sources. First, we know that the moon was personified as women (with breasts, obviously) and that the Greeks called breasts apples. Secondly, all indications are that man recognized the moon long before he did the sun and that even the rising and setting suns were first associated with the moon and night and only later with the day. Finally, we have every reason to believe that the ancients first believed that fertility was the exclusive property of females. There are hosts of myths in which mother goddesses give birth to offspring without first mating with a male—god or otherwise.

If early man did believe that females alone were fertile, then he would have been faced with an unpleasant prospect when the role of males in procreation finally was realized. Then he would have been forced to go back on a long-held belief. The compromise would have been the sharing of fertility, mythologized as a gift from female to male. In all likelihood, this would have been symbolized as a gift of fruit (for fruitfulness) from goddess to god. Perhaps this is one reason why the statues of Aphrodite show the love goddess proffering an apple.

Now in our reconstruction of the apple-goddess myth, we have Sun dying, then revived by eating one of Moon's apples; that is, by nursing at her breast. This is based on the part of the Hercules legend that claims that as an infant Hercules suckled at the breast of Hera, and it was this that made him immortal.

Therefore (symbolically), we have Sun gaining both immortality and fertility by the single gift of an apple, which is to say, by suckling at Moon's breast. Unhappily, this seems to lead to an inconsistency. One myth holds that Hercules achieved immortality at the end of his life by eating a sacred apple and submitting to his ritual slaughter. Now we say that Hercules gained immortality as an infant by suckling at Hera's breast.

This seeming paradox, that Hercules achieved immortality both at the beginning and at the end of his life, can be explained in this way: The apple played a dual role in the death and rebirth of the sun. First off, sunset symbolized the end of the sun king's reign, and sunset was a golden apple. This is one reason why the sun king was fed apples before his slaughter. However, sunrise was the symbol of the beginning of his successor's reign, and that also was a golden apple, a breast-apple—a suitable food for a king in the infancy of his reign. So the apple was the mediating symbol between the death and rebirth of the sun god.

The one constant figure in this drama was the moon. It was she who controlled the seasons, so it was she who determined the death and rebirth of the sun. For her, it was just a matter of which apple to give to whom—golden apple to the dying sun king, and breast-apple to his infant successor.

Finally in our creation myth, we have Moon and Sun as the parents of mankind. This is based on a myth that has Prometheus making the bodies of men and women out of water and clay and Athena breathing life into the clay figures. (Prometheus apparently was another sun god of the old dying-and-reviving stripe. He was a black-smith, with all the fiery implications attached to that idea, and he was the god who took the sun's fire and gave it to man. He was punished by Zeus for this act by being chained to a rock; he was later freed by yet another sun god, Hercules. Athena was a Libyan moon goddess accepted into the Greek pantheon, her real identity being carefully disguised. This creation story reveals her earlier role as one of the parents, with a sun god, of mankind.)

The Judgment of Paris

Now, let us return to the Judgment of Paris, which was first discussed in the preceding chapter. Literary sources identify the three goddesses in the story as Hera, Athena, and Aphrodite. Yet the three goddesses must be regarded as the moon shown under three guises. Can these two different sets of identities be reconciled? The answer is yes.

This ancient vase painting shows the Judgment of Paris as the Greeks imagined it circa B.C. 500. At left is Paris (identified on the vase by his other name, Alexander), then Hermes, the messenger god; the war goddess Athena; Hera, the queen of the gods; and the love goddess Aphrodite with winged attendants. From George Perrot and Charles Chipiez's *Histoire de l'Art dans l'Antiquite*, Paris, 1882–1914.

Hera, like all the goddesses in the Greek pantheon, was adopted into the Greek religion after the Greeks came down into the Mediterranean region. (The Aryans had worshipped only male deities.) Her name is equivalent to the Latin *Terra* (Mother Earth), and she was a reigning goddess on the Greek peninsula until she was transformed into the sister and wife of Zeus. However, even though Hera can be identified as Mother Earth, she also had attributes of a moon goddess. Even in classical Greece she was called "white-armed Hera," an allusion to moonbeams. And we know that in the city of Stymphalos, she was worshipped under three forms, as maiden, mother, and widow, which correspond to the three principal phases of the moon—new, full, and old. Really, there is nothing mythologically inconsistent with regarding her simultaneously as "Moon-Hera" and "Earth-Hera." In pre-Greek myth, the moon was regarded as the daughter without a father (hence a reincarnation) of Mother Earth.

Aphrodite originally was a goddess of the full moon, which in the pre-Greek lunar religions was the moon in her love goddess aspect. When Aphrodite was adopted into the Greek religion, she remained a love goddess but her lunar identity was concealed. Greek myth offers two versions of her birth: In one she was said to have sprung from the foam that surrounded the genitals of the god Uranus when they were thrown into the sea. (He had been castrated by his son Cronus.) The other story was that Aphrodite was born simply of sea foam, spontaneously, without a father or any male principal involved.

Fathered or not, however, Aphrodite sprang from the sea, an element of primordial earth. It also happens that the sea was the special domain of the moon goddess. She controlled the tides, after all, and the dark night sea reflected her face. In short, Aphrodite can be seen as a sister, daughter, or perhaps rebirth of the moon, in other words, the moon herself. Further, she is married to the blacksmith of the gods, whose gleaming furnace makes him a solar figure, and their match is the typical moon-sun linkup.

Athena also was born without a father. She was found beside a lake, born of water like Aphrodite, and nursed by three nymphs. When her worship reached Greece, the Greeks transformed her from a goddess without a father to a daughter of Zeus without a mother, having her spring full-grown from Zeus's head. She was accepted by the Greeks as a goddess of wisdom, which probably was one of her attributes back in Libya. There, however, her chief wisdom would have been that life is cyclical, like the phases of the moon, an idea the Greeks did not accept.

Assuming that Hera, Aphrodite, and Athena are moon goddesses and therefore would possess the apple as their own symbol, why would they assemble before Paris seeking the reward of an apple? The answer would seem to be that the Judgment of Paris was a later Greek reversal of an earlier myth in which the three goddesses (the moon in her three principal aspects) appeared to a prince to offer him apples. Each could offer him an apple in her own right. Hera could offer the apple of immortality, the breast that suckled the man-god Hercules; Aphrodite could offer the apple of life, fertility, and death; Athena could offer the apple of wisdom, the knowledge that all these things were so. To accept the moon's apple was to accept life in all its forms, and that included death. There was the promise of rebirth, of life after death. Faith can work wonders, but for most people, faith can only carry them so far.

A Christian Counterpart

Now, should anyone doubt that the apple was this kind of token of knowledge and immortality, there is a legend from the Christian era that preserves the myth quite perfectly. Saint Dorothea—a Christian, obviously—was being led off to execution, and the pagan lawyer Theophilus mockingly asked her to send him back some fruit from the paradise she believed would be her faith's reward.

"As you wish," Dorothea replied, asking the guards to allow her a moment to pray.

Suddenly, a boy appeared whom no one had seen approach. He carried a flower-filled basket containing three apples, fruit so beautiful they looked as though made from jewels. "Give these to Theophilus," Dorothea said, "and tell him there are more in Paradise where I hope to meet him."

Shortly after, Dorothea's head was lopped off. And Theophilus, who by now had tasted the apples, realized the meaning of Dorothea's convictions, namely that life on earth was simply a prelude to eternal life in Paradise and that Paradise was reached through the doorway of death. Theophilus embraced Christianity and soon afterward was himself martyred.

On the surface the story seems Christian enough, but if one knows a little Greek, there is something about the names Theophilus and Dorothea that doesn't ring quite true. Dorothea is formed out of two Greek words, *doron* and *thea*, meaning "gift of the goddess." Theophilus comes from *theos* and *philos*, meaning "lover of god." Whether there ever were two Christian martyrs named Dorothea and Theophilus is irrelevant, because this pious little story is not about them. It's the Christianization of a pagan apple story.

The historical St. Dorothea died circa A.D. 300, but the "gift of the goddess" in the story is the apple itself. And the "lover of God" (Theophilus) is the man who accepts the apples and eats them, gaining immortality and the knowledge of the meaning of life and death.

The Apple in Medicine: Moon Magic

If the apple held the promise of immortality, then surely it could restore a sick man to health. That, no doubt, was the sort of reasoning that made the apple an important plant in the medical lore of Europe. Some of the fruit's curative powers may have rubbed off from associations with such healing gods as Apollo, Hercules, and Dionysus, and all were regarded that way, but clearly, it was the apple-moon association, our apple goddess, that earned the fruit its reputation.

In 1597 John Gerarde wrote down what he called "the vertues" of the apple. Gerarde billed himself as a master in surgery, and his book was called *The Herball, or Generall Historie of Plantes*. Let these excerpts from Gerarde speak for themselves:

> Apples be good for an hot stomacke: those that are austere and somewhat harsh, do strengthen a weake and feeble stomacke proceeding of heate.
>
> Apples are also good for all inflammations or hot swellings, but especially for such as are in the beginning, if the same be outwardly applied. . . .
>
> There is likewise made an ointment with the pulpe of Apples and Swine grease and Rose water, which is used to beautifie the face and to take away the roughness of the skin, which is called in the shops *Pomatum*, of the Apples whereof it is made.
>
> The pulpe of the rosted Apples, in number fower or five, according to the greatnesse of the Apples, especiall of the Pome-water [a variety of apple], mixed in a wine quart of faire water, laboured togither untill it come to be as Apples and Ale, which we call Lambes Wool, and the whole quart drunke last at night, within the space of an hower, doth in one night cure those that pisse by droppes with great anguish and dolour; the strangurie [slow and painful urination], and all other diseases proceeding of the difficultie of making water; but in twise taking it never faileth any.

The list goes on, but clearly the point has been made. The apple was a potent medical agent once upon a time. Gerarde's learned con-

CARRYING TWO LARGE BASKETS OF WINTER APPLES
From S. E. Todd's *The Apple Culturist*, New York, 1871. (*Courtesy the Library of the New York Botanical Garden*)

temporary Sir Kenelm Digby, in his *Choice and Experimented Receipts In Physicke and Chirurgery*, even prescribed the apple in the treatment of breast cancer, recommending a hot plaster made of the pulp of roasted apples and hog's grease.

In a book called *The Queens Closet opened*, written in 1655 by someone signed W.M., we find a sweetened apple syrup recommended for melancholy.

This list clearly points to our basic association. All these various cures add up to nothing but moon magic. The apple is sacred to the moon; therefore, it is cold, and therefore it is good for treating "a weak and feeble stomacke proceeding of heate" and "all inflammations and hot swellings." The moon's face is smooth and beautiful; therefore, we find apples in an ointment "used to beautifie the face and take away the roughness of the skin." The moon controls the tides; therefore, we find apples in a mixture prescribed for "those that pisse by droppes with greate anguish and dolour." The moon's breasts are apples; therefore, we have apples recommended as "an anodyne cataplasm [soothing plaster] for cancered breasts." Finally, it is the moon that causes mental disorders (lunacy); therefore, we have a "syrup of pearmains [a variety of apple] good against melancholy."

Old beliefs die hard: An apple a day keeps the doctor away. Or we could take a second look at the wassail bowl. The bowl was round, a symbol of the sun's disk. The ale was heated, a reminder of the sun's life-giving warmth. There were roasted apples, a memento of the sun's summer heyday. And all were solar charms that the drinkers imagined would give the sun back his strength.

But it is also possible to give the apples a lunar meaning. Perhaps the midwinter revelers were, in effect, trying to give the sun some advice, acting out what the sun ought to do, miming the message: "Sun, just eat Moon's apples, and everything will be all right. You will feel like a new sun, and the crops will grow once more."

The Apple of Eden

Assuming that our path was lined with apple blossoms, not primroses, it seems safe to say we have found our apple goddess. She was the totem or protectress of some ancient tribe, the Apple tribe perhaps. At some point in her early career, she was identified as the moon goddess, and thereafter the apple was regarded as the moon's sacred token.

Later, when the patriarchal, sun-worshiping Aryans dominated Europe, the moon goddess's role in man's life was diminished but hardly eradicated. The apple remained a symbol of immense power.

We have gone full circle, and now we can return to the questions posed at the beginning of the previous chapter, questions that asked, in effect, "Why did the pagans of Europe consider the apple to be the fruit of Eden?" The answer is now close at hand.

The Bible said that Eve, the mother of all the living, gave the fruit to Adam. In pagan lore, the moon gave the apple to her husband, the sun, and they were the parents of mankind.

The Bible said the forbidden fruit was that of the Tree of the Knowledge of Good and Evil. In pagan lore, the apple was the fruit of the knowledge of another pair of opposites, life and death.

The second tree in Eden was the Tree of Life, meaning eternal life. In pagan lore, the apple held the same promise. It was the fruit that made a mortal man into a god.

6

Apples in Folklore:
Peels and Cores

This nineteenth-century bookplate shows the apple tree as it will grow when left to its own devices (without shaping or thinning). From J. F. Schreiber's *Bilder-Werke für den Anschauungs-Unterricht in Schule und Haus*, Esslingen, 1882.

In reconstructing the apple-goddess myth, we have omitted other material, stories and so on, that obviously came from later periods and did not reflect the early beliefs about the apple. Just the same, these interesting and amusing stories and customs do exist, and it would be a shame to let them pass unmentioned. So, these stories have been gathered together here—the peels and cores of apple love.

Apples of Love

The Norse sun god Freyr had fallen in love with a frost giant's daughter, Gerda, who was so beautiful that her white arms lit up both sea and sky. However, Freyr had not received permission from Odin, the chief of the gods, to marry Gerda, so he asked a go-between to court her in his stead. The go-between promised Gerda that she would receive Freyr's golden apples (he was also god of fruit and grain) if only she would say that Freyr was dearest to her. Gerda refused but relented when threatened with a life without the joys of love yet coupled with unquenchable desire.

Norse legend also contains the story of Rerir, who was childless and prayed to Odin and his wife Freyja, who was goddess of love, marriage, and death. The gods sent Ljod, in the form of a crow, with an apple, which she dropped in Rerir's lap. Later on, Ljod married Rerir's son, who had been conceived through the apple.

Apples of Immortality

Three Norse gods, Odin, Loki, and Hoenir, were out wandering, and food was hard to find. They came upon a herd of cattle, separated an animal from the herd, slaughtered it, and set about cooking it. After a respectable wait, the gods scattered the fire and tried the meat, but it was raw. They started the fire again, and the same thing happened. Then the gods heard a voice speaking from a nearby oak tree. It was the giant Thjazi, who had assumed the form of an eagle.

The giant said that he had prevented the meat from cooking but would relent if he were allowed to have his fill first. The gods agreed, but when the eagle swooped down, they attacked it. Loki stuck a pole in the eagle's back, and the great bird flew off with the pole still in him and Loki holding fast to the pole.

Loki thought he was going to be killed and pleaded with the eagle for his life. The eagle promised to consent if only Loki would lure the goddess Idunn, the guardian of the apples of immortality, into coming out of Asgard, the home of the gods, with her apples, which she kept in a chest. Loki got Idunn out of Asgard by telling her he had found, in a certain wood, apples that he thought were superior to hers. He also asked that she bring along her apples for comparison.

Instantly, Thjazi swooped down in his eagle plumage and carried off Idunn and her apples. It didn't take long for the gods to miss Idunn, because without her apples the gods swiftly grew hoary and old. Loki, under the threat of torture, donned hawk's plumage and flew to the giant's home. There Loki rescued Idunn by turning her into a nut and flying back to Asgard with the goddess grasped in his talons.

Connla, son of an Irish king named Conn, saw a stranger who told him she came from the Land of the Living, where people lived in perpetual feasting. The woman was invisible to everyone but Connla. Conn discovered his son in a seemingly one-way conversation and asked with whom he was speaking. The woman answered (and she could be heard by all) that she was one who faced neither death nor old age and that she loved Connla and wanted him to come with her.

The king summoned his druid (or priest) and ordered him to use his magic against the woman. The druid recited a charm to prevent Connla from seeing and all others from hearing the goddess, who left after giving Connla an apple.

The lovelorn Connla would eat nothing but the apple after that. Yet the apple never seemed to grow smaller. A month passed, and the goddess again appeared to Connla. This time she came in a glass boat, and again she invited him to join the immortals.

The king once more summoned his druid, but the goddess sang across the water to Connla that he would forget his grief at leaving his friends and that soon he would be in a land of joy where only women dwelled. Connla sprang into the glass boat, which sped across the sea, and never returned.

Several Irish heroes, Cuchulainne and Curoi among them, were attacking the stronghold of the god Midir, but their efforts were unsuccessful. Curoi promised to lead the heroes through the god's magical defenses if he could have the pick of the spoils. The heroes agreed, and Curoi lived up to his promise.

The spoils included Midir's daughter, Blathnat, whom Curoi chose as his prize, much to Cuchulainne's chagrin. Blathnat was not particularly pleased with the arrangement either. It seemed she would have preferred Cuchulainne, whom she asked to be her avenger.

Curoi, however, was immortal, and before vengeance could be accomplished, Blathnat had to learn the secret of his immortality. She extracted the secret from Curoi that his soul was contained in an apple, which could be found in the belly of a salmon that appeared in a spring on a hillside only once in seven years. He also told her that he would die only if the apple was split in half, something that could only be done with his own sword.

Blathnat gave this information to Cuchulainne, who waited seven years to catch the salmon and get the apple. Blathnat drew a bath for her husband, and when he was in it, she tied his long hair to the bedpost. Cuchulainne entered; Blathnat handed him her husband's sword. Cuchulainne cut the apple in two, then cut off Curoi's head.

The Persians tell of the holy man Anasindhu who was given the apple of immortality by the god Gauri as recognition of his wisdom and goodness. Anasindhu was about to eat the fruit, when he remembered that his wife had shared previous hardships and now deserved a share of the god's gift. She refused the offer, however, on grounds that she didn't care to be a holy man's wife, living in loneliness and poverty for the rest of time. She suggested selling the apple, which Anasindhu reluctantly agreed to do.

Anasindhu came out of the forest where he dwelt and went to the city. There he sold the apple to the king, who was about to eat the fruit when he started to reflect on life and decided that he was not worthy of the honor. He went into his garden to meditate and saw his wife there. She was the one who deserved the apple, he decided, because there was no one more beautiful in the world, and her beauty should delight the world forever.

The queen accepted the apple of immortality, and the king returned to the palace. Meanwhile, it had gotten dark, and before she began to eat the apple, she slipped off to a secluded spot in the garden for a rendezvous. By morning, the captain of the guard was the new possessor of the apple.

The captain was about to eat the apple when he remembered the servingmaid, whom he loved above all others. He decided he would make her immortal. However, the servingmaid felt she was unworthy, so after her beloved left, she took the apple to the king.

The king, of course, asked the maid who had given her the apple, and she replied that it was her betrothed, the captain of the guard. The captain was summoned; he confessed that the apple had been given to him by the queen, and he was executed immediately. The queen was burnt in the town square.

The king was so distraught over the entire affair that he decided to give away his riches, dress in rags, and live the life of a holy man. As he left the palace, Anasindhu, now dressed in finery, rode by on a litter. The king gave the apple back to Anasindhu, who decided that the gods must really have wanted him to be immortal. He went to bite the apple; his litter bearers stumbled, and the apple fell to the ground. A stray dog came loping by, snapped up the apple, and swallowed it in a single gulp.

The Germans tell of a young servant in a royal household who ate the flesh of a white snake and found he could understand the language of the animals. The very afternoon he ate the flesh he was threatened with prison after one of the queen's rings disappeared.

While pondering his bad luck, the young man overheard a duck remark that it had swallowed the ring that morning. The servant grabbed the duck and brought it to the cook, who prepared the bird for roasting. The cook found the ring in the duck's stomach, and the young man was exonerated.

The king, by way of apology, promised the young man any favor he should ask, and the young man asked to be released from the king's service and be allowed to seek his fortune. The request granted, the youth set out.

Along the way, he saved three fish who were stranded on dry land. He discovered that his horse had stepped on an anthill, and he led the animal away. He found three young ravens who had been thrown from their nest, and he killed his horse so they would have food.

Then he came to a land where the princess demanded that suitors carry out her wishes or lose their lives. The young man decided to pursue her hand, because she was very beautiful. When he confronted her, the princess threw a ring in the water and told him he must bring it back or die. The fish that the young man had saved returned his kindness and fetched the ring from the depths.

The princess then scattered ten sacks of millet and told the youth that he must pick every grain by morning or lose his life. During the night, the ants whose nest he had spared gathered all the millet. So the princess said she wanted an apple from the tree of life, and one of the three ravens the young man had saved by killing his horse brought the apple to him in its beak.

And so the young man became the princess' consort and, since her father was dead, the king of the entire country.

The Germans also tell of a young prince whose golden ball rolled into the cage of a wild man captured by the king's men. The boy begged the wild man to return the ball, and the latter said he would if the prince unlocked the cage with the key hidden under the queen's pillow. The boy freed the wild man, who then took the boy into the forest to save him from the king's wrath.

In the forest the wild man set the prince to guarding a spring-fed pool. He must prevent anything from falling into the pool and polluting it. One day the boy dipped his finger in the pool and it turned to gold. And the wild man, even though he was off in the forest at the time, knew what the prince had done. The next day one of the boy's hairs fell into the pool; it, too, turned to gold, and once more the wild man knew what had happened. The third day the prince was looking at his reflection in the pool, and his long hair fell over his shoulders, and every strand turned to gleaming gold. The pool was now polluted, the wild man said, and the boy could no longer stay in the forest.

Still, the wild man said he wished the boy no harm, because he knew the prince had not meant to pollute the pool. The wild man promised to help the prince if ever he was in need.

The prince then went out of the forest and found a castle, where he entered service as the gardener's assistant. The princess there took one look at the prince's golden hair and fell in love with him. She was convinced from the start that the boy was not a commoner. However, the prince did not want to reveal his identity, so he avoided the princess.

Then one day the prince's adopted kingdom was threatened by invasion, and the prince went to the wild man and asked for weapons and a horse so that he could join the fray. The wild man granted his request and provided a troop of iron soldiers as well. The young prince saved the day (and the kingdom), then returned everything he had borrowed to the wild man and resumed his disguise.

The princess wanted to learn the identity of the mystery knight who had saved her father's kingdom. She was certain it was the gardener's assistant. The king said he would hold a three-day-long feast at which the princess would throw a golden apple. Presumably, the mystery knight would come and catch the apple.

On the first night of the feast the princess threw the golden apple. The prince caught it but ran off without revealing his identity. The princess threw a second golden apple the second night, and the results were the same. The third night she threw a third golden apple, but this time the prince identified himself. So he and the princess were betrothed.

At the wedding feast the doors suddenly burst open and in marched a king with his courtiers. This king said that he had been bewitched and turned into the wild man but that the prince had broken the spell. The wild-man-turned-king then gave all his gold to the prince.

Prester John was a priest-king of immense power and wealth, was several centuries old, and his realm was somewhere in the East (or perhaps in Ethiopia). At least, that was the belief in much of Europe after the middle of the twelfth century, when a "letter" about this mythical figure made its first appearance on the Continent.

The palace of Prester John was supposedly built on the pattern of a castle that the Apostle Thomas built for the Indian king Gundoforus.

Ceilings, joists, and so on were made of shittah wood, the wood of the Ark of the Covenant. The gabled roof was of ebony, and at each end was a golden apple set with a carbuncle, a deep red jewel, so that the golden apples could shine by day and the carbuncles by night.

Lovers' Questions

These three examples come from Whitney and Bullock's *Folklore from Maryland*:

"Suspend some apples over a doorway, blindfold a girl with her hands behind her, and let her try to bite an apple. The one she bites will be her future husband."

"On Halloween, hang an apple on a string attached to a nail. The number of people passing under it will indicate the number of months that will elapse before you are married."

"In order to see one's future mate one need only stand before a mirror in a darkened room and eat an apple in front of the mirror at midnight. The face of the future husband will appear over the shoulder."

The following come from *Kentucky Superstitions*:

"If you can break an apple in two, you can get anyone you choose as your life partner."

"If you can break an apple in two after someone has named it, the one named loves you."

"As many fresh apple seeds as stick to the forehead when pressed against it, so many days it will be until you see your sweetheart."

"Shoot an apple seed up; your sweetheart lives in the direction that it goes."

"Name apple seeds and shoot them at the ceiling. The one that hits the ceiling shows the one that loves you the most."

"Put five apple seeds on your face and name them. The first to fall shows which one you will marry."

"Place apple seeds on a grate; name them. The ones that jump show which ones love you."

"Count apple seeds to learn the number of children you will have."

"If you can eat a crab apple without frowning, you can get the person you desire."

"On Halloween name an apple that hangs by a string. If you can bite it, you are beloved by the person that you have named."

"On Halloween, he who can bite an apple by bobbing for it will marry."

These two beliefs also can be found in *Kentucky Superstitions*, but they are not properly called divination rituals. The first is a love charm, the second a sign or omen.

"If you hold an apple in your armpit until it is warm and then eat it, your sweetheart will love you."

"If you find twin peaches, apples, cucumbers, or the like, you will be married soon."

Nebraska was also fertile ground for apple lore, as these examples, taken from Margaret Cannell's "Signs, Omens and Portents in Nebraska Folklore" show:

"Have someone snap and name your apple for some man friend. When the apple is eaten count the seeds to divine your future relationship, thus,

> One I love, two I love, three I love I say;
> Four I love with all my heart, five I cast away;
> Six he loves, seven she loves, eight they both love;
> Nine he comes, ten he tarries;
> Eleven he courts and twelve he marries;
> Thirteen they quarrel, fourteen they part;
> Fifteen they die with a broken heart.

"Name two apple seeds. Put them on a hot stove. If they hop together, the people for whom they were named will marry."

Assorted Beliefs

From *Folklore from Maryland*:

"If your fruit trees blossom twice in a year, you will have ill luck, and still worse if they bear fruit three times."

"If a fruit tree blossoms twice in a year, the owner will lose a friend by death."

"Dream of plucking fruit out of season and you'll quarrel without reason."

"Peeling peaches, pears, or apples alone gives bad luck. If you wish your sweetmeats to turn out well, get a friend to help you."

"To find water for your well, hold an apple twig by the forked ends, with the palms of the hands upwards, and it turns when you come to the right spot."

"Throw an apple on the roof of your house, and if it falls off, you will be happy."

From *Kentucky Superstitions*:

"If you bite an apple off to use as bait in a dead fall [trap], you will catch a rabbit."

"If you can break an apple in half with your hands, you will always be your own boss."

Apple Cures

This charm was recommended in nineteenth-century New England as a cure for ague, a type of recurring fever (often malaria) accompanied by chills:

"The patient in this case is to take a string made of woolen yarn, of three colours, and to go by himself to an apple tree; There he is to tie his left hand loosely with the right to the tree by the tri-coloured string, then to slip his hand out of the knot, and run into the house without looking behind him."

This cure, described in a book called *Folk Medicine* by William George Black, was based on the idea of magical transference. The ritual was supposed to transfer the disease to the tree. The wart cures below, both found in *Kentucky Superstitions*, were based on the same concept:

"You can make a wart go away by burying an apple secretly and saying this charm: 'As this apple decays, so let my warts go away.'"

"Make an incision in the bark of an apple tree. When the place grows over, your wart will be gone."

The business of burying the apple, mentioned in the first wart cure, was a humane refinement over earlier versions of this type of cure. It used to be that one would rub the wart with something like a stone, then leave the stone at the crossroads. The stone was merely a go-between in transferring the wart to some hapless passerby.

The following is recorded in Kinelm Digby's *Choice and Experimented Receipts in Physicke and Chirurgery*:

"An anodyne Cataplasm [soothing plaster] for cancered Brests, the first Cataplasm that Mr. Bressieus applyed to Mrs. Brents cancered Brest when it begun to break, was this,

"Take an old mellow Pippin, cut off the Cap at the top of it, and then take out the core leaving the sides of the Apple whole, that the melted Grease may not get out: fill that hole with Hogs-grease, then

Nine apples drawn from life by Johann Hermann Knoop. From Knoop's
Pomologie, Amsterdam, 1771. (*Courtesy the Library of the New York Botanical
Garden*)

cover it with the Cap, and set the Apple to roast, when it is well roasted to pap (by which the Hogs-grease will be imbibed into the substance of the Apple) take it and pare off all the pairing, and break and mingle all the pap, that it may spread well, and be an uniform Pulp: Spread it thick on Linnen, and lay it warm upon the sore, putting the Bladder [covering] over it. This is an excellent Cataplasm, to take away an cool, and dissolve the swelling and hardness if it be dissolvable; and if not, to make it break and separate with ease and without sharpness. This is to be changed every twelve to twenty hours, according as it groweth dry."

According to *The Queens Closet opened* by W. M., published in 1655, "The Lady Gorings remedy for a sharp urine" was this:
"Boyl running water with Liquorish till be something strong of it, boyl also in it a Pippin or two, when it is boyled, put in some brown Sugar candy, drink of it every morning fasting a pretty draught."

According to *The American Frugal Housewife*, by Mrs. (Lydia Maria) Child, published in 1833, "The following poultice for the throat distemper, has been much approved in England: —The pulp of a roasted apple, mixed with an ounce of tobacco, the whole wet with spirits of wine, or any other high spirits, spread on a linen rag, and bound upon the throat at any period of the disorder."

Digby's *Choice Receipts* gives the following cure for diarrhea·
"For the greatest flux or looseness, take a right Pomwater the greatest, or two little ones, roast them very tender to pap, take away the skin and the coar, and use only the pap, and the like quantity of chalk finely scraped; mix them together before the fire, and work them well to a plaister, then spread it upon a linnen cloath warmed very hot as may be suffered and so bind it to the Navel 24 hours; use it two or three times till the Flux stays off."

Perhaps the basis of this cure was in the notion of fighting fire with fire. Certain apples could cause diarrhea, so they could have been thought useful in curing it. The idea of putting snow on frostbitten limbs is the same sort of thing, and centuries ago an entire body of medical practices, called homeopathy, was based on the principle.

However, the apples in the cure are roasted, while raw apples, especially unripe fruit, are the chief offenders. If roasting the apples was thought to reverse their basic qualities, then the cure would be based on a concept of opposites, called allopathy. John Gerarde wrote this in his book *The Herball* in 1597:

"Rosted Apples are alwaies better than the rawe, the harme whereof is both mended by the fire, and may also be corrected by adding unto them seedes and spices."

Finally, the cure could have been based on lunar magic, since the moon, as mentioned previously, had control over water. Today the refined version of these cures is found in the form of the antidiarrheal Kaopectate, whose chief ingredient is pectin.

From Mrs. Child's *American Frugal Housewife*, comes this recipe:
"A spoonful of ashes stirred in cider is good to prevent sickness at the stomach. Physicians frequently order it in cases of cholera-morbus."

The Greek Herbal of Dioscorides ... (*A.D. 512*), *Englished by John Goodyear A.D. 1655* imparted the following medical information:
"The leaues & ye blossomes & the sprigs of all sorts of Apple trees ar binding, especially of ye Quince tree, & ye fruit being unripe is binding, but being ripe, not soe; but those apples which are ripe in ye spring tyme, breed choller, being hurtful of all that is sinewy and flatulent."

Gervase Markham's *The English Hus-wife*, published in 1615, contained a cure for jaundice:

"For the yellow Iaundisse, take two penny-worth of the best English Saffron, drie it and grinde it to an exceeding fine powder, then mix it with the pap of a rosted apple, and give it to the diseased party to swallow down in the manner of a pill; and do thus divers mornings together, and without doubt it is the most present cure that can be for the same, as hath been often proued."

Markham also suggested the following cure:

"For wormes in the belly, either of childe or man, take aloes cicketrine [a purgative], as much as halfe a hazell Nut, and wrap it in the pappe of a roasted apple, and so let the offended party swallow it down in the manner of a pill fasting in the morning."

The following cure is found in Gerarde's *The Herball*:

"Apples cut in peeces, and distilled with a quantitie of of Camphere and butter milke, taketh away the marks and scars gotten by the small pockes, being washed therewith when they grow unto their state or ripeness: provided that you give unto the patient a little milke and saffron, or milke and mithridate [a toxic herb related to aconite] to drinke to expell to the extreme parts that venome which may lie hid, and as yet not seene."

The two apple cures given below come from *The Queens Closet opened*:

"Mr. Lumley's Pippin drink for a Consumption:

"Take the thick paring of six Pippins, boyl them in three pintes of Spring water, then sweeten them with Sugarcandy, whereof drink the quantity of a Wine glass when you go to bed. In a feaver it is very good with a little Syrup of Limons."

"Syrup of Pearmains good against Melancholy

"Take one pound of the juyce of Pearmains, boil it with a soft fire, till half be consumed, then put it in a glass, and there let it stand till it be settled, and put to it as much of the juice of the leaves and roots of

Borage [an herb with a cucumberlike flavor], Sugar half a pound, Syrup of Citrons three ounces, let them boyl together to the consistence of a Syrup."

Americana

John Chapman was born in Leominster, Massachusetts, in 1774. He moved to the Ohio–Indiana frontier sometime around the beginning of the nineteenth century. He was a land speculator with holdings of more than 1,100 acres, and a trader. He was a convert to the religion of the Swedish mystic Emanuel Swedenborg and was married to a Choctaw Indian woman who died of malaria.

Chapman was known as an eccentric but pleasant fellow, with a good capacity for liquor and a way with animals and plants. He also went around planting dog fennel, an herb he thought was a cure for malaria, and apple seeds, which he obtained from cider mills.

Unhappily, the dog fennel Chapman sowed broadcast was an obnoxious weed, and the so-called orchards he planted were of little consequence because apple trees grown from seed seldom produce fruit of any great value. However, through the imagination of people like Rosella Rice, who as a youngster knew the man and grew up to write about him in a most sentimental way, John Chapman was transformed from an admittedly colorful figure into an American saint— Johnny Appleseed.

Rosella Rice's "Recollections," which first appeared in 1888 in the *History of the Ashland County Pioneer Historical Society*, form the basis of much of the myth of Johnny Appleseed. They apparently were written down sometime during the 1850s and appeared in the publications of various Ohio historical societies during the 1870s and 1880s, the years surrounding the centennial of American independence and a time of great sentimental interest in American history.

"He [Johnny] was a man rather above middle stature, wore his hair and beard long and dressed oddly. He generally wore old clothes that he had taken in exchange for the one commodity in which he dealt—apple trees. . . . Sometimes he carried a bag or two of seeds on an old horse, but more frequently he bore them on his back, going from place to place on the wild frontier, clearing a little patch, surrounding it with a rude enclosure and planting seeds therein. He had little nurseries all through Ohio, Pennsylvania and Indiana. If a man wanted trees and was not able to pay for them, Johnny took his note, and if the man ever got able and was willing to pay the debt, he took the money thankfully; but if not, it was well. . . . He was such a good, kind, generous man, that he thought it wrong to expend money on clothes to be worn just for their fine appearance. He thought if he was comfortably clad, and in attire that suited the weather, it was sufficient. . . .

"All the orchards in the white settlements came from the nurseries of Johnny's planting. Even now all these years, and though this region [Ohio] is densely populated, I can count from my window no less than five orchards or remains of orchards that were once trees taken from his nurseries. . . .

"One of his nurseries is near us, and I often go to the secluded spot on the quiet banks of the creek shut in by sycamore trees, with the sod never broken since the poor man did it. And when I look up and see the wide out-stretched branches over the place like outspread arms in loving benediction, I say in a reverent whisper 'Oh the angels did commune with the good old man, whose loving heart prompted him to go about doing good. . . .'

"On the subject of apples he was very charmingly enthusiastic. One would be astonished at his beautiful description of excellent fruit. I saw him once at the table when I was very small, telling about some apples that were new to us. His prescription was poetical, the language remarkably well chosen. . . . I stood back of mother's chair, amazed, delighted, bewildered, and vaguely realizing the wonderful powers of true oratory. . . .

"He was never known to hunt any animal or to give any animal pain; not even a snake. One time when overtaken by night while traveling he crawled into a hollow log and slept till morning. In the other end of the log was a bear and her cubs. Johnny said he knew the bear would not hurt him, and that there was room enough for all.

"The Indians all liked and treated him very kindly. They regarded him from his habits as a man above his fellows. He could endure pain like an Indian warrior; could thrust pins into his flesh without tremor. Indeed so insensible was he to acute pain that treatment of a wound or sore, was to sear it with hot iron and then treat it as a burn. . . .

"In 1838 he resolved to go further on. Civilization was making the wilderness to bloom like a rose. Villages were springing up, stagecoaches laden with travellers were common, schools were everywhere, mail facilities were very good, frame and brick houses were taking the place of the humble cabins. . . . This must have been a sad task for the old man, who was then well stricken in years, and one would have thought that he would have preferred to die among his friends. He came back two or three times to see us all in the intervening years that he lived; the last time was in the year he died, 1845. In the Spring of that year, one day after traveling twenty miles, he entered the house of an acquaintance in Allen County, Indiana, and was as usual, cordially received. He declined to eat anything except some bread and milk, which he ate, sitting on the doorstep occasionally looking out on the setting sun.

"Before bedtime he read from his little books (a Bible and two or three works by Swedenborg that he carried with him) 'fresh news from heaven,' and at the usual hour for retiring he lay down upon the floor, as was his invariable custom. In the morning the beautiful sight supernal was on his countenance, the death angel had touched him in the silence and the darkness, and though the dear old man essayed to speak, he was so near dead that his tongue refused its office. The physician came and pronounced him dying, but remarked that he never saw a man so perfectly calm and placid, and he inquired particularly concerning Johnny's religion. His bruised and bleeding feet, worn, walk the gold paved streets of the New Jerusalem, while we so brokenly and crudely narrate the sketch of his life. A life full of labor and pain and un-selfishness, humble unto self-abnegation, his memory glowing in our hearts, while his deeds live anew every springtime in the fragrance of the apple-blossoms he loved so well."

In his *Myths and Legends of Flowers, Trees, Fruits, and Plants,* Charles M. Skinner wrote that toward the end of the seventeenth century in Franklin, Connecticut, there lived a farmer named Micah Rood, who was considered something of a no-account by his neighbors. One day, a wandering peddler was found dead on Rood's farm. The peddler's head had been deeply gashed and his pack emptied. The suspicion naturally fell on Rood, but there was no real evidence that he actually had committed the crime.

After the episode, however, Rood became a recluse, neglecting his farm and letting house and outbuildings fall to ruin. Whether it was a case of guilty conscience or disgust with his neighbors and with life was never established. Then something strange happened.

The peddler had died beneath an apple tree that bore yellow fruit, but in the summer after the murder, the tree began to bear red-skinned apples. What is more, in each fruit the flesh surrounding the core was stained a bright red. And the town gossips were quick to offer the explanation that the change in the fruit and the decline that followed Micah Rood to his grave were the results of the dead man's curse.

Roger Williams is one of the most famous advocates of civil and religious freedom in the history of America. He arrived in the New World in 1630, and by 1636 his libertarian views had earned him the honor of being banned in Boston, actually banished from the entire Massachusetts Bay Colony. With a band of followers he fled beyond the boundaries of English jurisdiction to what is today Rhode Island, which was free from any English claims or patents.

He and his companions founded the city of Providence, and after many years and many struggles he died at the age of eighty-four and was buried on his estate, which was near the city he founded. Later on, the city had grown to the point that Williams's gravesite had become an obstacle to progress, so it was decided to move his body. When the grave was opened, according to Katherine Stanley Nicholson's *Historic American Trees*, the workmen discovered that an apple tree had sent its roots into Williams's grave and that the roots had traced the outlines of his body.

Famous Trees

In *The White Goddess*, Robert Graves mentions a tree famous in Irish lore, the Oak of Mugna, which the Strong Upholder (perhaps a god) endowed with three fruits: the acorn, the apple, and the hazelnut. The tree was said to be thirty cubits (45 feet) around and its crown was said to cover an entire plain. It was always covered with leaves, and as soon as one fruit fell, it was replaced by another.

According to *Historic American Trees*, during the American Revolution, a band of Hessian soldiers reportedly planted an orchard about three miles north of Winchester, Virginia. Sixteen of the trees, which were of the French Fameuse variety, survived into the twentieth century and were still bearing fruit in the late 1930s.

One of the most famous apple trees in American lore is the Miami Apple Tree, which stood opposite the city of Fort Wayne, Indiana. This description has been excerpted from *Harper's New Monthly Magazine* of April 1862:

"At the junction of the St. Mary and St. Joseph rivers, where they form the Maumee River, or Miami of the Lakes, in Indiana, is a rich plain. . . . It is opposite the city of Fort Wayne, that stands upon the site of the Indian village of Ke-ki-on-ga. There was once one of the most noted villages of the Miami tribe of Indians; and there Mish-i-ki-nak-wa, or Little Turtle, the famous Miami chief, was born and lived until late in life. He and his people have long since passed away, and only a single living thing remains with which they were associated. It is a venerable Apple-Tree, still bearing fruit when . . . visited . . . late in September, 1860. It is from a seed doubtless dropped by some French priest or trader in early times. It was a fruit-bearing tree a hundred years ago, when Pe-she-wa (Wild Cat) or Richardville, the successor of

The Miami Apple Tree stood opposite the city of Fort Wayne, Indiana, and marked the site of one of the principal villages of the Miami Indians. The tree was bearing fruit a century before this engraving was made in 1862. From *Harper's New Monthly Magazine*, April 1862.

Little Turtle was born under it; and it exhibits now—with a trunk more than twenty feet in diameter, seamed and scarred by age, and the elements—remarkable vigor.

"Glimpses of the city of Fort Wayne may be seen from the old Apple-Tree; and around it are clustered scenes from the close of the last century, when American armies were sent into that region to chastise the hostile Indians. On the Maumee, near by, a detachment of General Harmar's troops were defeated and decimated by the Indians, under Little Turtle, in 1791. The sanguinary scene was at the ford, just below the Miami village; and Richardville, who was with Little Turtle, always declared that the bodies of the white men lay so thick in the stream that a man could walk over on them without wetting his feet."

Part II

RECIPES

GATHERING APPLES
An engraving from *The Dollar Monthly Magazine*, December 1864.

Most of the recipes in this book have been drawn from cookbooks written before this century. Many date back several centuries, and one, Diced Pork and Apples (see recipe on page 180), goes back at least a millennium and a half. When reconstructing recipes of such great age, a certain amount of guesswork and interpretation is involved. Old cookery books are notoriously incomplete in their details. One is never 100 percent certain what "Take some flour . . ." really means.

The texts of most of the old recipes have been quoted in their entirety to provide a flavor of what the old cookbooks were like. In a sense, by quoting the old recipe and providing an updated version, the book becomes a guide on how to go about interpreting old cookbooks. Further, if you are not happy with the updated version, you have the option of going back to the original and reinterpreting it for yourself.

For example, the majority of the updated recipes contain far less spice than the original versions. The cooks and diners of centuries ago went in for highly seasoned dishes. Grating half a nutmeg into a pie was hardly unusual a few centuries back. That's about 1½ teaspoons of spice. Most people today would be satisfied with ¼ teaspoon of nutmeg in the same dish.

Also in keeping with modern tastes and uses, butter or vegetable shortening has been used on occasion when the original recipe called for lard or suet. Again, you have the option of reverting to the earlier version.

All of the recipes have been chosen with twentieth-century tastes in mind. Nearly every ingredient should be available in a supermarket. The few exceptions include rosewater and orange-flower water, which are available in specialty and gourmet shops; venison; and a spice called galingale, which doesn't seem to be available at all. However, the fourteenth-century recipe for Apple Blossom Caudle (see page 207), which calls for galingale, features three other spices, so the loss isn't really profound.

The "Right" Apple

Cookbooks prate on about which apple is right for which dish, and the answers, for the most part, add up to a lot of bunk. Tart apples usually are specified for cooking, yet the sweet Golden Delicious apple is as fine as any in an apple pie and makes a good sauce as well.

Summer apples, such as the Lodi and the Yellow Transparent, are branded as cooking apples, yet if you like a tart apple with a mild but pleasant flavor, you will find either of these varieties delightful to eat out of hand.

Apple producers would have you believe that the Rome Beauty is one of the best of all baking apples, and although a good Rome is an acceptable apple, many of the Romes in the marketplace are mealy and practically tasteless. This apple's only saving grace is that it usually keeps its shape when baked.

Even Delicious, which is identified as strictly a dessert fruit, can be used in cookery, although it tends to get mushy. Delicious will make a mild but pleasant applesauce, and it's the right apple to use if you are wild about that Delicious flavor. Just go easy on the spices, or they will be all you will taste.

Clearly, the right apple for you is the apple you like best. If you like a Gravenstein apple, then Gravensteins make the best pies, but so do Rhode Island Greenings, Jonathans, Northern Spies, and Baldwins, depending on who's doing the eating. There are some differences in how the various apples perform in cooking, and these can be pointed out. But the final choice should depend as much on taste preferences as on any concept of ideal performance. If you still can't make up your mind, here is a chart that shows how the "experts" usually stack up apple varieties.

	Puddings and Dessert	Pastry	Sauce	General Cooking	Salads*
Baldwin	x	x		x	
Cortland	x			x	x
Delicious	x				
Early McIntosh	x		x	x	
Golden Delicious**	x	x	x	x	x
Grimes Golden	x		x	x	
Gravenstein	x	x	x	x	
Jonathan	x	x		x	
Lodi		x	x	x	
McIntosh	x	x	x	x	
Newtown Pippin	x			x	
Northern Spy	x	x		x	
R. I. Greening		x	x	x	
Rome Beauty		x		x	
Stayman Winesap	x		x	x	
Wealthy	x	x	x	x	
Winesap	x	x	x	x	
Yellow Transparent			x	x	
York Imperial		x	x	x	

    * Cortland and Golden Delicious are recommended for use in salads because their flesh stays whiter longer after cutting.

    ** When using Golden Delicious in cooking, remember to cut back on the sugar specified in the recipe. Golden Delicious is considerably sweeter than most other apples recommended for cooking.

7

Salads

SPANISH SALAD
HAM AND APPLE SALAD
CELERY AND APPLE SALAD
CUPID SALAD
HORSERADISH AND APPLE RELISH

Spanish Salad

Cut into small pieces 1 cup of celery, 3 cooked beets, 2 cold potatoes, and ½ a raw apple. To this add the boneless part of 6 sardines. After removing the skin, reduce to a pulp. Mix all ingredients together thoroughly, pour over French dressing and serve on lettuce.

From *The Palisades Cook Book*,
by the Ladies Aid Society of
the Tenafly, New Jersey,
Presbyterian Church, 1910

6   sardines, skinned and boned
1   cup chopped celery
3   cooked beets, chopped
2   large boiled potatoes, peeled and chopped
½   apple, peeled, cored, and chopped
½   cup French dressing
     Lettuce leaves

Mash the sardines and add them to the chopped vegetables and apples in a large mixing bowl. Add the dressing and toss all together. Serve on beds of lettuce leaves.

SERVES 4

Ham and Apple Salad

⅔   cup mayonnaise
1   tablespoon cider vinegar
½   teaspoon salt
¼   teaspoon dry mustard powder
¼   teaspoon white pepper
2   large stalks celery
2   apples peeled, cored, and diced
2   cups diced cold cooked ham
     Lettuce leaves

Combine mayonnaise, vinegar, salt, mustard powder, and white pepper together in a small bowl. Blend well, and chill while preparing other ingredients.

Julienne the celery by cutting the stalks into thin strips, then cutting the strips crosswise in ½-inch-long pieces.

Combine celery, apples, and ham in a mixing bowl. Pour the dressing over all and toss lightly. Serve on beds of lettuce leaves.

SERVES 4

Celery and Apple Salad

Two cups of celery chopped fine, grated rind of 1 orange, 1 cup of apple cut into dice. Mix the above with the following mayonnaise: 1 very cold egg yolk, with 2 drops of vinegar on it and 1 teaspoonful of onion juice and yolk of 1 boiled egg, 1 cup of cold olive oil, a tablespoon of sugar, a tablespoon of lemon juice, a little salt and cayenne pepper and a half teaspoonful of mustard. Mix thoroughly by stirring oil drop by drop to the egg, and a few drops of vinegar, lemon, salt, pepper, etc., which have previously been thoroughly mixed together. Serve either on white lettuce or in the apple cups, with cheese balls rolled in the chopped walnuts.

From *The Palisades Cook Book,*
by the Ladies Aid Society of
the Tenafly, New Jersey, Presbyterian Church, 1910

The ambitious might want to make their own mayonnaise using the recipe above, but a good facsimile of the dressing has been provided below.

Dressing:
1   cup prepared mayonnaise
1   tablespoon lemon juice
1   teaspoon onion juice
½   teaspoon dry mustard
    Pinch of cayenne

Cheese Balls:
¼   cup grated sharp Cheddar cheese
¼   cup cream cheese
1   tablespoon butter
1   tablespoon dry red wine
¼   cup finely chopped walnut meats
2   cups chopped celery
1   cup chopped apples
    Peel of 1 orange, grated
    Lettuce leaves

To make dressing, combine mayonnaise, lemon juice, onion juice, dry mustard, and cayenne together in a small bowl. Beat well, and chill for 10 to 15 minutes.

To make cheese balls, combine Cheddar cheese, cream cheese, butter, and wine in a bowl and work into a smooth, stiff paste. Form into 12 small balls and roll them in the chopped walnut meats. Set aside.

Combine celery, apples, and grated orange peel in a mixing bowl. Add dressing, and toss.

Serve salad on beds of lettuce leaves, garnished with the cheese balls.

SERVES 6

Cupid Salad

4   oranges
2   bananas
1–3 cups of sugar
1   pint strawberries
1   large tart apple
1   egg
1   tablespoonful of brandy

Cut the oranges in half, scoop out the pulp, keeping the peel intact. Slice the bananas, hull and slice the strawberries. Place all the materials on ice. Make a dressing of the apple, sugar and brandy. Grate the apple and sprinkle it with sugar as you grate so as to keep it from turning dark; add to it the brandy and un-beaten white of egg, and with a wire egg beater beat until it is stiff and fluffy. With sharp scissors cut the orange cups near the top into scallops, and tie them together in pairs with baby ribbon. When ready to serve fill the orange cups with the prepared fruit and heap the dressing on top.

From *The Palisades Cook Book,*
by the Ladies Aid Society of
the Tenafly, New Jersey,
Presbyterian Church, 1910

4   oranges
2   *large, ripe bananas, peeled and sliced*
1   *pint fresh strawberries, hulled and sliced in halves*

Dressing:
1   *large apple, peeled, cored, and quartered*
⅓   *cup sugar*
1   *egg white*
1   *tablespoon brandy*

Cut the oranges in half and scoop out the pulp, keeping the peels intact. Reserve peels for use as serving cups.

Chop orange pulp coarsely and combine with bananas and strawberries in a covered bowl. Refrigerate while preparing the dressing.

To prepare dressing, grate apple quarters, sprinkling a bit of sugar over the pulp several times to keep it from turning dark. Combine pulp, remaining sugar, egg white, and brandy in a small mixing bowl. Beat mixture with a wire whisk until stiff and fluffy. Refrigerate, covered, while preparing the orange peel cups.

Scallop cut edges of orange peel halves with a pair of sharp scissors. Put two halves on each serving plate and tie together with a piece of narrow ribbon. Fill with fruit mixture and heap dressing on top.   SERVES 4

Horseradish and Apple Relish

Peel and grate a tart apple. To two parts grated apple add one part grated horseradish. Serve with cold meats.

From *The Palisades Cook Book,*
by the Ladies Aid Society of
the Tenafly, New Jersey, Presbyterian Church, 1910

This recipe needs no explaining, since it adds up to a milder sauce or relish for the usual horseradish applications, such as in the accompaniment of cold roasted or boiled meats or cold poached fish.

YIELDS APPROXIMATELY 1 CUP

8

Main Dishes

BAKED CHICKEN WITH APPLES

SAUCE FOR ROAST RABBIT

DEVONSHIRE SQUAB PIE

APPLE AND RAISIN STUFFING FOR POULTRY

APPLE AND SAUERKRAUT GOOSE STUFFING

VENISON CUTLETS WITH APPLES

CHESHIRE PORK PIE

PORK ROYAL

DICED PORK AND MATIAN APPLES

PORK AND APPLE FRITTERS

SAUSAGES AND APPLES

SCHNITZ UND KNEPP (APPLES AND DUMPLINGS)

APPLE OMELET

Baked Chicken with Apples

First season them with cloves and mace, pepper and salt, and put to them Currance and Barberies, and slit a Napple and cast Sinnamom and Suger upon the Apple, and lay it in the bottome, and to it put a dish of Butter, and when it is almost enough baked, put a little Sugar, Vergious and Orenges.

From *The Good Husvvifes Ievvell,*
by Thomas Dawson, London, 1587

Dawson's recipe specifies "vergious," which is known today as verjuice and is the sour, unfermented juice of unripe grapes or crab apples. If verjuice is available, it can be substituted for the wine used in the updated recipe.

2 *apples, peeled, cored, and sliced thick*
¼ *cup seedless raisins*
4 *tablespoons sugar*
¼ *teaspoon ground cinnamon*
Juice and grated peel of ½ lemon
2 *tablespoons butter*
1 *2- to 3-pound chicken, split in half*
1 *teaspoon salt*
¼ *teaspoon black pepper*
¼ *teaspoon ground mace*
⅛ *teaspoon ground cloves*
¼ *cup dry white wine*
Juice and grated peel of ½ orange

Preheat oven to 325 degrees F.

Line bottom of a buttered baking dish with apple slices. Sprinkle raisins, 2 tablespoons of the sugar, cinnamon, lemon juice, and grated lemon peel over apple slices. Dot with butter.

Sprinkle both sides of each chicken half with a mixture of the salt, black pepper, mace, and cloves. Place chicken, skin side up, in the baking dish, atop the apples.

Bake uncovered in preheated oven for about 2 hours.

About 45 minutes before chicken is done, combine wine, remaining 2 tablespoons of sugar, orange juice, and grated orange peel in a small saucepan. Bring to a boil, reduce heat, and simmer for 10 to 15 minutes. Pour this mixture over the chicken, as a glaze, and resume baking.

Garnish chicken with cooked fruit and serve.

SERVES 4

Sauce for Roast Rabbit

To dresse a hare wash her in faire water, perboile her, then lay her in cold water, then lard her and rost her, and for sauce take red wine, salt, vinegar, ginger, pepper, cloves and mace, put these together, then mince onions and apples and frye them in a panne, then put your sauce to them with a little sugar, and so let them boile together, and then serve it.

From *The Good Husvvifes Ievvell*,
by Thomas Dawson, London, 1587

Although this sauce was originally prepared for rabbit, it can also be a piquant accompaniment to roast turkey, duck, or goose.

1   *large apple, cored and diced*
1   *large yellow onion, peeled and chopped*
2   *tablespoons butter*
2   *cups dry red wine*
2   *tablespoons cider vinegar*
1   *teaspoon sugar*
½   *teaspoon salt*
¼   *teaspoon ground ginger*
¼   *teaspoon freshly ground black pepper*
¼   *teaspoon ground mace*
     Pinch of ground cloves

In a large skillet or saucepan, sauté apple and onion in butter until onion is transparent. Add wine, cider vinegar, sugar, salt, ginger, black pepper, mace, and cloves. Stew all together over a low flame for about 5 minutes.

Serve the meat in the fashion of the day by lining a platter with thick slices of crusty bread, heaping the carved meat atop the bread, and pouring the sauce over all.

YIELDS 3 CUPS

Devonshire Squab Pie

Cover your dish with a good crust, and put at the bottom
of it a layer of sliced pippins, and then a layer of mutton steaks,
cut from the loin, well seasoned with pepper and salt. Then put
another layer of pippins, peel some onions, slice them thin, and
put a layer of them over the pippins. Then put a layer of mutton
and then pippins and onions. Pour in a pint of water, close up the
pie, and send it to the oven.

From *Modern Domestic Cookery,*
by W. A. Henderson,
New York, 1857

Like the Cheshire Pork Pie (page 176), this recipe tastes best when the meat has been browned, something that is heresy, no doubt, to the residents of Britain. Since mutton is seldom available in America, lamb has been substituted.

1   *pound lean lamb*
    Vegetable oil as needed
3   *large yellow onions, peeled and sliced*
    Pie Pastry for a two-crust pie (recipe on page 232)
3   *large apples, peeled, cored, and sliced thick*
1   *teaspoon salt*
½   *teaspoon black pepper*
½   *cup water*

Preheat oven to 425 degrees F.

Cut lamb into slices about ¼ inch thick. Brown slices well in a skillet, adding a spoonful or two of vegetable oil if necessary. Set aside.

Brown onions in fat left in the skillet.

Line a 9-inch pie pan with approximately half of the pastry dough.

Put in a layer of apple slices, then a layer of lamb. Sprinkle in half the salt and pepper. Then add another layer of apples, a layer of onions, and another layer of lamb. Sprinkle in the rest of the salt and pepper, and finish with a layer of apples and a layer of onions.

Add the water and cover the pie pan with top portion of pastry dough. Seal, trim, and flute edges of crust, slashing top in several places to let steam escape.

Bake at 425 degrees F. for 15 minutes, then reduce heat to 350 degrees F. and bake for 35 to 40 minutes more.

Serve hot.

YIELDS 1 9-INCH PIE

Apple and Raisin Stuffing for Poultry

3   *tart apples, peeled, cored, and chopped*
½   *cup seedless raisins*
5   *cups of cubed soft bread*
¼   *cup sugar*
    Juice and grated peel of ½ lemon
1   *teaspoon ground cinnamon*
½   *teaspoon salt*
¼   *cup sweet cider*

Combine all ingredients in a mixing bowl. Toss all together well.

Although the quantities of ingredients will stuff a 5- to 6-pound bird, you can double the recipe for a 10- to 12-pound bird. If using recipe to stuff a chicken or turkey, add 2 tablespoons melted butter. Goose and duck are fatty enough not to require this addition.

YIELDS ENOUGH STUFFING
FOR A 5- TO 6-POUND BIRD

Apple and Sauerkraut Goose Stuffing

8   *cups sauerkraut*
1   *yellow onion, peeled and chopped*
2   *tablespoons butter*
8   *large apples, peeled, cored, and coarsely chopped*
4   *juniper berries, crushed (if available) or 2 teaspoons caraway seeds*

Wash sauerkraut in fresh water if it seems excessively sour.

In a small skillet sauté onion in butter until golden brown.

In a mixing bowl combine sauerkraut, onions, apple slices, and juniper berries. Mix well.

YIELDS ENOUGH STUFFING
FOR A 10- TO 12-POUND GOOSE

Venison Cutlets with Apples

Wipe, core, and cut four apples in one-fourth inch slices. Sprinkle with powdered sugar, and add one-third cup port wine; cover, and let stand thirty minutes. Drain, and sauté in butter. Cut a slice of venison one-half inch thick in cutlets. Sprinkle with salt and pepper, and cook three or four minutes in a hot chafing-dish, using just enough butter to prevent sticking. Remove from dish; then melt three tablespoons butter, add wine drained from apples, and twelve candied cherries cut in halves. Reheat cutlets in sauce, and serve with apples.

From *The Boston Cooking-School Cook Book,*
by Fannie Merritt Farmer,
Boston, 1896

This recipe specified a chafing dish because gas and electric stoves had yet to become standard household fixtures at the time *The Boston Cooking-School Cook Book* was first published. This was an easy supper dish that did not require firing up a huge coal range late in the day. A chafing dish can still be used, if the flame is hot enough, but most cooks today will find a frying pan and a modern gas or electric range much more convenient. When venison is unavailable, a beefsteak may be substituted.

4	*apples, cored and sliced ¼-inch thick*
¼	*cup sugar*
⅓	*cup port*
6	*tablespoons (approximately) butter*
1½	*pounds venison or beefsteak cutlets*
	Salt and pepper to taste
12	*candied (glacé) cherries, cut in halves*

Put apple slices in a small covered bowl and sprinkle sugar and 2 tablespoons of port over them. Set aside for 30 minutes. Then drain slices, reserving liquid, and sauté in 2 tablespoons butter until just tender, about 10 minutes. Arrange slices on a warm serving platter and set aside.

Season meat with salt and pepper to taste and fry in a hot skillet for 1 to 2 minutes on each side, using just enough butter to keep meat from sticking to pan. Remove meat, reduce heat, and melt remaining butter in pan. Then add the remaining port and the candied cherries. Return meat to skillet and cook over moderate flame until reheated, about 4 to 5 minutes. Add meat to serving platter with apples and pour the sauce over all.

SERVES 4

Cheshire Pork Pie

Cut two or three pounds of lean fresh pork into strips as long and as wide as your middle finger. Line a buttered dish with puff-paste; put in a layer of pork seasoned with pepper, salt, and nutmeg or mace; next a layer of juicy apples, sliced and covered with about an ounce of white sugar; then more pork, and so on until you are ready for the paste cover, then pour in half a pint of sweet cider or wine, and stick bits of butter all over the top. Cover with a thick lid of puff-paste, cut a slit in the top, brush over with beaten egg, and bake an hour and a half.

This is an English dish, and is famous in the region from which it takes its name. It is much liked by those who have tried it, and is considered by some to be equal to our mince-pie.

From *Common Sense in the Household,*
by Marion Harland, New York, 1884

Although Mrs. Harland's recipe did not specify browning the meat, there are definite advantages in doing so. For one thing, it improves the flavor of the pie. Second, it cuts down the baking time, so the apples are not reduced to mush. The updated version of this recipe halves the original.

1 to 1½	*pounds lean fresh pork*
	Vegetable oil as needed
	Pie Pastry for a two-crust pie (recipe on page 232)
1	*teaspoon salt*
½	*teaspoon black pepper*
½	*teaspoon ground mace*
3 to 4	*apples, peeled, cored, and sliced thick*
2	*tablespoons sugar*
½	*cup sweet cider*
1	*tablespoon butter*

Preheat oven to 425 degrees F.

Cut pork into ¼-inch-thick slices. Brown well in a skillet, adding a teaspoonful or two of vegetable oil if necessary to prevent sticking.

Line a 9-inch pie pan with approximately half the pastry dough. Put in a layer of the pork and sprinkle over it half the salt, pepper, and mace. Add a layer of apple slices and sprinkle with 1 tablespoon sugar. Add a second layer of pork and remaining seasonings and finish with a layer of apples, topping with remaining 1 tablespoon of sugar.

Add cider and dot pie with butter.

Top with balance of pastry dough. Seal, trim, and flute edges of crust, slashing top in several places to let steam escape.

Bake at 425 degrees F. for 15 minutes; then reduce heat to 350 degrees F., and bake for 35 to 40 minutes more.

Serve hot.

YIELDS 1 9-INCH PIE

Pork Royal

Take a piece of shoulder of fresh pork, fill with grated bread and the crust soaked, pepper, salt, onion, sage and thyme: a bit of butter and lard. Place in a pan with some water; when about half done, place around it some large apples; when done, place your pork on a dish, with the apples round it; put flour and water on your pan, flour browned, some thyme and sage; boil, strain through a very small colander over your pork and apples.

From *Housekeeping in Old Virginia*,
by Marion Cabell Tyree,
Louisville, Kentucky, 1879

1   *3- to 5-pound pork shoulder roast*
2   *tablespoons lard*

Stuffing:

  2   *tablespoons butter*
  4   *cups stale bread cubes*
  1   *medium yellow onion, peeled and chopped*
  2   *teaspoons salt*
1¼ *teaspoons black pepper*
 ½ *teaspoon sage*
 ½ *teaspoon thyme*
1½ *cups water*
  6   *large apples*

Gravy:

 ¼ *cup flour*
  1   *cup water*
      Generous pinch of sage
      Pinch of thyme

Have butcher remove bones from roast. Sew the roast, leaving one side open to form a pocket for the stuffing.

Preheat oven to 350 degrees F.

Melt lard and butter together in a saucepan.

Combine in a mixing bowl bread cubes, onion, 1 teaspoon salt, ¼ teaspoon black pepper, ½ teaspoon sage, and ½ teaspoon thyme. Pour melted lard and butter over the bread-cube mixture and add ½ cup hot water. Toss all together lightly.

Fill pocket of roast with stuffing and sew up opening. Sprinkle 1 teaspoon salt and 1 teaspoon black pepper over meat.

Place roast in an uncovered roasting pan, add 1 cup water, and roast in a preheated oven, allowing 35 to 40 minutes per pound of meat.

Core apples and peel them about a third of the way down from the stem end. About 45 minutes before serving, put apples around meat in roasting pan.

When roast is done, put it on a platter and surround with baked apples.

To make gravy, put roasting pan on top of the stove. Add flour to juices in pan and brown over moderate heat. Gradually stir in 1 cup water, and cook, stirring constantly, till gravy is thickened. Add pinches of sage and thyme and cook for 5 minutes more.

Strain gravy and pour over meat and apples. Serve hot.

SERVES 6 TO 8

STUFFING VARIATIONS

Apple Stuffing: Replace 1 cup bread cubes in recipe above with 1 cup chopped apples. Increase lard and butter to 3 tablespoons each.

Prune and Apple Stuffing:

10	prunes
	Water to cover prunes
2	tablespoons butter
1	cup bread crumbs
3	large apples, peeled, cored, and coarsely chopped
1	teaspoon sugar
	Peel of ½ lemon, grated
½	teaspoon ground cinnamon

Cover prunes with water and cook in a saucepan until tender, about 40 minutes. Cut each in half and remove pits.

Melt butter in a second saucepan, then add cooked prunes, bread crumbs, and chopped apples. Toss.

Add sugar, grated lemon peel, and cinnamon, and toss once more.

Diced Pork and Matian Apples

In a small pot put oil, liquamen, broth; cut up a leek, cori-
ander, small meatballs. Cut into cubes a piece of roast pork with
the skin left on. Put them together to cook. Halfway through the
cooking, add Matian apples, cores removed, and cut into dice.
While this is cooking, grind together pepper, cumin, coriander
(fresh or seed), mint, asafetida root; pour over this vinegar, honey,
liquamen, a little defritum and some of the cooking liquid, adding
vinegar to taste. Put up to boil and when it is boiling, add some
crumbled pastry and thicken the sauce with this. Sprinkle with
pepper and serve.

From *De Re Coquinaria* ("On Cookery"),
by Coelius Apicius,
first to fourth century A.D.

The author of *De Re Coquinaria*, which is probably the oldest surviving cookbook, is sometimes identified as the gourmet Apicius, who lived during the reign of Tiberius Caesar (A.D. 14 to 31). However, the language of the surviving texts dates from about the fourth century. Whether the versions we have are later editions of an earlier work or the writings of a fourth-century Roman, also named Apicius, is a question that remains unanswered.

Liquamen, also called *garum*, was a sauce prepared from dried salt fish and from the earliest days was prepared outside the home by professional sauce makers. *Liquamen* provided the salt in Roman recipes and can be replaced best by the modern soy sauce of the Orient. *Defritum* was wine cooked down to about a third of its original volume; a heavy, sweet port makes an acceptable substitute in the modern kitchen. Asafetida, known botanically as *Ferula asafoetida*, is a plant related to the carrot. Its flavor is rather like that of garlic, which has been substituted in the following updated recipe. Fresh coriander can be obtained in many Spanish groceries, where it is known as *cilantro*. The meatballs in the original version have been omitted.

2 tablespoons olive oil
2 tablespoons soy sauce
1 cup chicken broth (1 chicken bouillon cube in 1 cup boiling
    water)
1 leek, coarsely chopped
2 tablespoons minced fresh coriander
2 cups small cubes of cold roast pork
4 medium apples, peeled, cored, and cut into eighths

Sauce:
¼ teaspoon black pepper
¼ teaspoon ground cumin
1 tablespoon minced fresh coriander
1 tablespoon minced fresh mint
2 cloves garlic, minced
¼ cup cider vinegar
1 tablespoon honey
1 tablespoon soy sauce
¼ cup port
2 tablespoons fine bread crumbs

In a large saucepan combine olive oil, soy sauce, chicken broth, chopped leek, minced coriander, and cubed pork. Bring to a boil, then simmer for 10 minutes.

Add apples and simmer until apples are just tender, about 20 to 25 minutes more.

Meanwhile, prepare the sauce in a small saucepan by combining the black pepper, cumin, coriander, mint, garlic, cider vinegar, honey, soy sauce, and port. Bring to a boil, then simmer for 15 minutes.

Add bread crumbs to the sauce and allow mixture to thicken. Thin with cooking liquid from the pork-apple mixture as needed until the sauce has the consistency of heavy cream.

Spoon pork-apple mixture into individual serving dishes and top with sauce.

SERVES 4

Pork and Apple Fritters

Prepare a light batter, freshen or use cold boiled or baked pork; cut it fine enough for hash, and fry it a little to extract some of the fat for frying the fritters. Peel sour apples, and cut or chop them not quite as fine as the pork; mix first the pork and then the apples in the batter, and fry them brown.

From *Godey's Lady's Book*,
Philadelphia, May 1867

1   cup flour, sifted
1   teaspoon baking powder
¾   teaspoon salt
¼   teaspoon black pepper
2   egg yolks, lightly beaten
1   cup finely diced cold cooked pork
1   cup chopped, peeled, and cored apples
2   egg whites, beaten stiff but not dry
    Deep fat for frying

In a mixing bowl, sift together flour, baking powder, salt, and pepper. Add beaten egg yolks, diced pork, and chopped apples. Mix well.

Fold in beaten egg whites, and fry large spoonfuls of the mixture in hot fat, which should cover the bottom of a heavy skillet to a depth of at least ½ inch. Brown on one side, then turn once to brown the other side.

Drain on paper towels before serving.

YIELDS ABOUT 8 FRITTERS

Sausages and Apples

The mode of frying sausages is so simple, and generally known that it needs no description. However, we shall notice one way of which the cook may not be informed. Take six apples, and slice four and take the cores clean out. Fry the slices with the sausages till they are of a nice light brown colour. When done put the sausages into the middle of the dish, and the apples round them. Garnish with the apples quartered.

From *Modern Domestic Cookery*,
by W. A. Henderson,
New York, 1857

6   apples
1   *pound country sausage links*

Peel, core, and slice 4 of the apples and set aside.

Prick sausage skins with a fork to keep them from bursting. Place in a large skillet over moderate heat. As soon as sausage fat covers the bottom of the skillet, add apple slices. Turn sausages several times to brown on all sides. Turn apples once to brown both sides.

Quarter and core the 2 remaining apples. When sausages and apples are nicely browned, drain on paper towels and arrange on a serving platter. Garnish with raw apple quarters.

SERVES 4

Schnitz und Knepp
(Apples and Dumplings)

Take of sweet dried apples (dried with the skins on, if you can get them) about one quart. Put them in the bottom of a porcelain or tin-lined boiler with a cover. Take a nice piece of smoked ham washed very clean, and lay on top; add enough water to cook them nicely. About twenty minutes before dishing up, add the following dumplings. Dumplings.—Mix a cup of warm milk with one egg, a little salt, and a little yeast, and enough flour to make a sponge. When light, work into a loaf. Let stand twenty minutes before dinner, then cut off slices or lumps, and lay on the apples, and let steam through.

From *Godey's Lady's Book,*
Philadelphia, June 1866

4   cups dried apple slices (see page 217 for information on dry-
      ing apples)
1   4- to 5-pound shank end smoked ham
½   cup dark brown sugar

Dough:
4   cups flour, sifted
2   tablespoons baking powder
2   teaspoons salt
1   teaspoon black pepper
¼   cup butter, softened
3   eggs
1   cup milk

Wash apples, place in a large bowl, cover with cold water, and soak overnight.

Wash ham with fresh water and wipe dry. Put ham and brown sugar in a large pot that has a tight-fitting cover. Add enough water to cover the ham. Simmer uncovered for 2 hours.

Add apples, plus the water in which they were soaked, to the ham and simmer, uncovered, for ½ hour more.

To prepare the dough, sift together in a mixing bowl the flour, baking powder, salt, and black pepper. Set aside.

In another large mixing bowl, cream the softened butter, and beat in the eggs one at a time. Beat in the milk, then add the flour mixture, about 1 cup at a time, beating well after each addition.

Drop dough into the cooking liquid in heaping tablespoonfuls.

Cover the pot tightly and simmer for 20 to 25 minutes more, until dumplings are done.

Serve the ham on a platter, surrounded by the schnitz (apples) and the knepp (dumplings).

SERVES 8 TO 10

Apple Omelet

Stew the apples, when you have pared and cored them, as for apple-sauce. Beat them very smooth while hot, adding the butter, sugar, and nutmeg. When perfectly cold, put with the eggs, which should be whipped light, yolks and whites separately. Put in the yolks first, then the rosewater, lastly the whites, and pour into a deep bake-dish, which has been warmed and buttered. Bake in a moderate oven until it is delicately browned. Eat warm—not hot—for tea, with Graham bread. It is better for children—I say nothing of their elders—than cake and preserves.

From *Common Sense in the Household,*
by Marion Harland, New York, 1884

This sweetened fruit omelet is rather like a soufflé. Mrs. Harland's version, however, was not original. It is almost a word-for-word copy of one found in the November 1871 issue of *The Lady's Friend,* a magazine published in Philadelphia.

6 *large apples, peeled, cored, and sliced*
¼ *cup water*
1 *tablespoon butter*
5 *teaspoons sugar*
  Nutmeg to taste
8 *eggs, separated*
1 *teaspoon rosewater*

In a large saucepan, stew the apples in ¼ cup water until they have reached a saucelike consistency, about 20 to 25 minutes. Beat them, gradually adding the butter, sugar, and nutmeg. Allow to cool.

Preheat the oven to 350 degrees F.

In a mixing bowl, beat the egg yolks until they are light and lemon colored. In a separate mixing bowl, beat the egg whites until they are stiff but not dry.

Fold the beaten egg yolks, the rosewater, and, finally, the stiffly beaten egg whites into the cooled apple mixture.

Butter a large soufflé pan or any deep dish, preferably one with vertical sides, and pour in the mixture. Bake in a preheated oven until firm, about 45 minutes. Check for "doneness" by sticking a thin-bladed knife into the center of the omelet; if the blade comes out dry, the omelet is ready.

SERVES 6 TO 8

9

Side Dishes

Fritters

The three fritter recipes included here provide a mini-lesson in cooking history regarding leavening. In the fourteenth-century recipe for Apple Fritters (page 188) no leavening is specified. The batter is only one step removed from the most primitive kind, which would contain only flour and water. However, the recipe does contain eggs, which act as a leavening of sorts.

When the fritter batter hits the hot fat, two things happen: The heat causes steam to form and starts to cook the eggs; and as the eggs cook and become hard, they catch little bubbles of steam. The batter fills up with the bubbles and becomes light and fluffy.

The seventeenth-century recipe (page 191) calls for eggs and yeast, a natural leavening agent. The yeast, which is made up of countless tiny one-celled plants, feeds upon the starches and natural sugars in the batter, then secretes carbon dioxide gas. This gas puffs up the batter, and when the batter is fried, the eggs catch both gas and steam bubbles, making the batter even fluffier.

The nineteenth-century recipe (page 192) specifies eggs, sour milk, and baking powder. The lactic acid in the sour milk reacts upon the baking powder (sodium bicarbonate) to produce the same harmless carbon dioxide gas the yeast formed. Sour milk and baking soda is no more effective as a leavening agent than yeast but acts much faster. It takes an hour or two for the yeast to produce enough carbon dioxide to fluff up the batter. The chemical reaction between the soda and the acid is almost instantaneous. (The yeast does give a definite "yeasty" flavor to the batter, while soda is practically tasteless when used in moderation.)

Since the late nineteenth century, baking powder has been the principal type of chemical leaven, although soda is still specified in certain recipes, such as those using fruit juices, which contain natural acids. Baking powder has a combination of ingredients which produce carbon dioxide gas whether or not there are any acidic ingredients in the recipe.

Fourteenth-Century Apple Fritters

For to make Fruturs: Nym flowre and eyryn and grynd peper and safronn and mak therto a batour and par aplyn and kyt hem to brode penys and cast hem theryn and fry hem in batour wyth fresch grees and serve it forthe.

From *The Forme of Cury* (circa 1390),
edited by Samuel Pegge, London, 1780

Preparing apple preserves—A kitchen scene from the seventeenth century. From *The French Gardiner*, translated by John Evelyn, London, 1672.
(*Courtesy the Library of the New York Botanical Garden*)

The language of this recipe is Middle English, spoken and written in England from about 1125 to about 1475. In modern English, the recipe might read something like this:

> *To make fritters, take flour and eggs and ground pepper and saffron and make of them a batter. And pare apples and cut them into broad pennies, then cast them into the mixture and fry them in the batter with fresh grease. And serve the dish forth.*

Batter:
2	eggs, well beaten
3	tablespoons flour
¼	teaspoon black pepper
	Generous pinch of saffron
	Pinch of salt

2 to 3	apples
	Vegetable shortening or lard for frying

To make batter, combine eggs, flour, black pepper, saffron, and salt, and beat together well in a mixing bowl. Set aside.

Core the apples whole, then peel them. Slice apples into rings, cutting them fairly thick, so that they won't break in frying.

Beat the batter once more, then dip each apple ring into the mixture. Allow the excess batter to drip from each slice before frying in a fairly hot skillet with several tablespoons of melted vegetable shortening or lard. Fry fritters to a golden brown on each side. Drain on paper towels and serve.

SERVES 4

Seventeenth-Century Apple Fritters

To Make Fritters: Take half a pint of Sack, a pint of Ale, some Ale yeast, nine Eggs, yelks and whites, beat them very well, the eggs first then altogether, put in some Ginger, and salt, and fine flower, then let it stand an hour or two, then shred in the Apples when you are ready to fry them, your suet must be all Beefe suet or halfe Beef, and half Hogges suet tryed out of the lease.

From *The Queens Closet opened,*
by W. M.,
London, 1655

The size of this recipe reflects the social conditions of the mid-seventeenth century, when cookbooks were aimed at well-to-do women (most poor people couldn't read) who had households that included many servants, guests, and relatives. "Sack," of course, is dry sherry; "yelks" are egg yolks; "flower" is flour, and "suet tryed out of the lease" is suet that has been melted and strained to remove the lecs (sediment).

Batter:
- ½   *cup ale*
- ½   *teaspoon dry active yeast*
- ¼   *cup dry sherry*
- ½   *teaspoon ground ginger*
- ¼   *teaspoon salt*
- 2   *eggs*
- 1   *cup flour*

2 to 3   *large apples, peeled, cored, and thickly sliced*
       Deep fat for frying

Warm the ale gently in a saucepan, and soften the yeast in it for about 10 minutes.

In a mixing bowl, beat together the dry sherry, ginger, salt, and eggs, and add the mixture to the ale.

Stir in flour gradually. Batter should be thick enough to coat apple slices generously. If batter is too thin, add more flour; if too thick, add more ale. Cover mixing bowl and set aside for 2 hours to let yeast raise the batter.

Dip apple slices in batter and fry in hot fat (365 degrees F.) for 2 to 3 minutes, or until fritters are golden brown. Drain on paper towels and serve.

SERVES 4

Nineteenth-Century Apple Fritters

Sour milk, 1 pt.; saleratus, 1 tea-spoon, flour to make a batter not very stiff; 6 apples, pared and cored; 3 eggs.

Dissolve the saleratus in the milk; beat the eggs, and put in; then the flour to make a soft batter; chop the apples to about the size of small peas, and mix them well in the batter. Fry them in lard, as you would dough-nuts. Eaten with butter and sugar.

From *Dr. Chase's Recipes,*
by A. W. Chase,
Ann Arbor, Michigan, 1874

Saleratus is an old-fashioned name for baking soda, which reacts with the lactic acid in the sour milk (or buttermilk) to leaven the fritter batter. If neither sour milk nor buttermilk are available, replace with the same quantity of sweet milk and substitute 1 teaspoon of baking powder for the baking soda.

Batter:
1 1/2 cups sifted flour
  1/4 teaspoon salt
  1/2 teaspoon baking soda
    2 tablespoons sugar (optional)

1   *egg*
¾   *cup sour milk, or buttermilk*
3   *large apples, peeled, cored, and chopped*

      Deep fat for frying
2   *tablespoons butter, or to taste*
      Sugar to taste

In a mixing bowl, sift together the flour, salt, baking soda, and sugar.

In another, large mixing bowl, beat the egg well and beat the milk into it. Add the flour mixture to this liquid and beat well.

Stir in the chopped apples, and fry large spoonfuls of the mixture in hot fat, which should cover the bottom of a skillet to a depth of at least ½ inch. Brown fritters on both sides and drain on paper towels.

Top with bits of butter and a sprinkling of sugar and serve.

SERVES 8

Apple, Raisin, and Nut Rissoles
(Rissoles on a Fish Day)

Item, commonly they be made of figs, raisins, roast apples and nuts peeled to counterfeit pine-kernels and powder of spices; and let the paste be well saffroned and let them be fried in oil. And if binding be necessary, amidon binds and rice too.

From *Le Menagier de Paris* (circa 1393),
translated by Elizabeth Power as
The Goodman of Paris,
London, 1928

Miss Power's translation is carefully contrived to sound like an early English cookbook. Unhappily, her translation requires nearly as much explaining as a sixteenth-century work might. "Pine-kernels" are pine nuts, the *pignoli* of Italian cookery. "Powder of spices" means ground spice or spices. "Paste" is pastry, while *amidon* is starch, in French. The slightly bitter, aromatic flavor of the saffron in the pastry contrasts nicely with the sweet, fruity filling.

Saffron Tea:
        Generous pinch of saffron
 ½   cup water
        Pie Pastry for a two-crust pie (recipe on page 232)

Filling:
 1   cup Applesauce (recipe on page 200), sweetened and spiced to
          taste
 ½   cup chopped seedless raisins
 ½   cup chopped blanched walnut meats
 2   teaspoons cornstarch

        Deep fat for frying

To prepare the Saffron Tea, combine saffron and water in a small saucepan. Place over heat and bring to a boil. Remove from heat, cool, then strain. Use this saffron tea to make the pastry for the rissoles by moistening the pastry dough with 4 to 6 tablespoons of it.

To prepare the filling, combine Applesauce, chopped raisins, chopped walnuts, and cornstarch in a mixing bowl.

Roll out flavored pastry dough to a thickness of ⅛ inch, and cut into 12 3-inch squares.

Put about a tablespoon of filling into the center of each square and fold corner to corner, forming triangles. Seal edges well, moistening them slightly with water if necessary.

Fry rissoles in moderately hot deep fat (360 to 370 degrees F.) until they are golden brown, turning once to brown the second side. Drain on paper towels and serve.

YIELDS 12 RISSOLES

Fried Apples

Take any nice sour cooking apples, and, after wiping them, cut into slices one-fourth of an inch thick; having a frying-pan ready, in which there is a small amount of lard, say ½ to ¾ of an inch in depth. The lard must be hot before the slices of apple are put in. Let one side of them fry until brown; then turn, and put a small quantity of sugar on the browned side of each slice. By the time the other side is browned, the sugar will be melted and spread over the whole surface.

Serve them up hot, and you will have a dish good enough for kings and queens, or any poor man's breakfast, and I think that the President would not refuse a few slices, if properly cooked.

From *Dr. Chase's Recipes*,
by A. W. Chase,
Ann Arbor, Michigan, 1874

The apples should be cored before slicing, but whether they should be peeled is a matter of choice. If the skins are tender, which is often the case with summer apples, for example, peeling is unnecessary.

ALLOW ONE APPLE PER SERVING

Boiled Apple Dumplings

Take half a dozen of the largest apples you can get, pare them, and take out the cores with an apple corer, fill the holes with quince or orange marmalade, or with beaten cinnamon and lemon-peel shred fine, mixed with powdered sugar; rub half a pound of butter with a pound of flour; make it into a stiff paste with cold water, roll a piece out round, put in the apple and close the paste over it. Tie them in separate cloths, and boil them one hour; then carefully turn them into a dish, sprinkle powdered sugar over them, with pats or slices of butter and powder sugar in plates.

From *The New Art of Cookery,*
by Richard Briggs,
Philadelphia, 1792

When Briggs mentions "powdered sugar," he is talking about our granulated sugar. In Briggs's day, sugar came in hard, cone-shaped loaves, which had to be broken into bits and powdered with a mortar and pestle before use.

½   *cup sugar*
½   *teaspoon ground cinnamon*
    Peel of ¼ lemon, shredded fine
4   *large apples, peeled and cored*

Crust:
2   *cups sifted flour*
½   *teaspoon salt*
½   *cup butter*
4 to 6   *tablespoons ice water*

Butter and sugar to taste

Combine sugar and cinnamon. Put this mixture and a few shreds of lemon peel into the cavity of each cored apple. Set apples aside.

To prepare the crust, sift together the flour and salt in a mixing bowl. Cut in the butter with a pastry blender or a pair of knives until it is cut into small bits the size of small peas. Add ice water a tablespoon at a time, tossing the dough with a pair of forks after each addition. Keep adding water until dough holds together when lightly squeezed.

Divide dough into 4 equal portions and roll each portion out to a circle about 9 inches in diameter. Place an apple over the center of each circle of dough and wrap the dough up the sides of the apple, sealing it at the top.

Prepare 4 clean squares of cloth, each about 10 to 12 inches on a side (four men's handkerchiefs will do nicely), by dipping the squares in hot water, wringing them out, and thoroughly dusting them with flour. Place an apple in the middle of each cloth, wrap the cloth around the apple, and tie it tight with a piece of string.

Place the dumplings into a pot of rapidly boiling water and cover the pot. Boil for 45 minutes, then remove them from the pot and plunge them into cold water for a few seconds. Untie the strings and gently peel the cloth from the dumplings.

Top each dumpling with a pat of butter and a spoonful or two of sugar. Serve hot. SERVES 4

Red Cabbage and Apples

1   *medium-sized red cabbage, coarsely shredded*
3   *large apples, peeled, cored, and sliced thick*
3   *large yellow onions, peeled and sliced thick*
     Salt and black pepper to taste
¼   *pound butter*
1   *cup dry red wine*

Arrange the shredded cabbage, apple slices, and onion slices in alternating layers in a large covered saucepan, sprinkling a little salt and pepper over each layer.

Cut the butter into 6 or 8 pats and place on top of the vegetables.

Add the wine, cover saucepan tightly, and let stew over very low heat for 2½ to 3 hours. SERVES 8

Squash and Apple Pudding

Core, boil and skin a good squash, and bruize it well; take
6 large apples, pared, cored, and stewed tender, mix together;
add 6 or 7 spoonsful of dry bread or biscuit, rendered fine as meal,
half pint milk or cream, 2 spoons of rose-water, 2 do. wine, 5 or
6 eggs beaten and strained, nutmeg, salt and sugar to your taste,
one spoon flour, beat all smartly together, bake.

From *American Cookery*,
by Amelia Simmons,
Hartford, Connecticut, 1796

This recipe can be made with any winter squash, such as Hubbard, butternut, or pumpkin.

1	*3½- to 4-pound winter squash*
6	*large apples, peeled, cored, and sliced thick*
¼	*cup water*
¼	*cup melted butter*
1	*cup milk or light cream*
1½	*cups sugar*
1	*teaspoon salt*
6	*eggs, well beaten*
½	*cup bread crumbs*
¼	*cup flour*
2	*teaspoons rosewater*
¼	*cup port or sweet sherry*
½	*teaspoon grated nutmeg*
	Whipped cream (optional)

Cut squash in half and scoop out seeds and strings. Cut into smaller pieces, and peel. Finally, cut into little chunks, about 2 inches on a side. Steam in a strainer over boiling water until squash is tender, about ½ hour. Mash squash in a bowl. Return to strainer and drain if squash seems particularly wet and runny.

Combine apple slices and water in a covered saucepan. Place over high heat and bring to a boil. Reduce heat and simmer until apples are tender, about ½ hour. Mash apples well.

Preheat oven to 350 degrees F.

Measure out 4 cups of cooked squash and 2 cups of cooked apples and combine in a large mixing bowl. Stir in the butter and mix well. Beat in the milk, sugar, salt, and beaten eggs. Add bread crumbs and flour and mix well. Add rosewater, port, and nutmeg and beat all together well.

Pour into a buttered, 3-quart baking dish and bake in preheated oven until custard is set, about 1¼ to 1½ hours. Test custard by sticking the blade of a knife into the center; if blade comes out clean, the custard is done. If custard starts to brown too quickly, loosely cover top of baking dish with a sheet of aluminum foil.

Serve warm as a side dish at dinner, or cold, topped with whipped cream, as a dessert.

SERVES 12

Apples and Sauerkraut

4 *cups sauerkraut*
4 *large apples, peeled, cored, and sliced ¼ inch thick*
2 *juniper berries, crushed (or 2 teaspoons caraway seeds)*
3 *tablespoons butter*

Wash sauerkraut in fresh water if it seems excessively sour.

Arrange sauerkraut and apples in alternate layers (about 2 or 3) in a large saucepan. Add crushed juniper berries and water to barely cover. (If caraway seeds are used, sprinkle some of the seeds over each layer of apples and sauerkraut.)

Cover saucepan and simmer until apples are tender, about 25 minutes. Remove cover from saucepan, add butter, and cook over high heat until most of the water has evaporated, watching pot closely lest it scorch.

Place in a serving dish and serve immediately.

SERVES 5

Applesauce

Pare and core six large apples, cut them in quarters, put them in a stew pan, with a little water to keep them from burning, a bit of cinnamon and lemon-peel, cover them close, and stew them gently till tender; take out the cinnamon and lemon-peel, bruise them well with a wooden spoon, put in some moist sugar and a little butter, and stir well till the butter is melted.

From *The New Art of Cookery*,
by Richard Briggs,
Philadelphia, 1792

6	*large apples, peeled, cored, and quartered*
½	*cup water*
1	*cinnamon stick*
	Peel of ¼ lemon
½	*cup sugar*
1	*tablespoon butter*

Combine apples, water, cinnamon stick, and lemon peel in a saucepan. Bring to a boil. Reduce heat, cover, and simmer for 20 minutes.

Remove lemon peel and cinnamon, mash apples well, and stir in sugar and butter.

Serve hot with roast pork or poultry.

YIELDS ABOUT 2 CUPS

Hot Applesauce (Appulmose)
for Meat and Fish

Nym appelyn and seth hem and lat hem kele and make hem throw a cloth and on flesch dayes kast therto god fat breyt of Bef and god wyte grees and sugar and safronn and almonde mylke of fische dayes, oyle de olyve and gode powdres and serve it forthe.

From *The Forme of Cury* (circa 1390),
edited by Samuel Pegge,
London, 1780

This Middle English recipe might read something like this in modern English:

Take apples and poach them. And let them cool and put them through a strainer. And on flesh days, add good, rich beef broth and good white grease and sugar and saffron. On fish days, add almond milk, olive oil and ground spices. And serve it forth.

This recipe is actually two recipes in one. The first mixture would be used as a sauce with meat, the second with fish. The "god wyte grees" and the "oyle de olyve" have been omitted to lighten the flavor of the sauces.

SAUCE FOR MEAT

3   *large apples, peeled, cored, and sliced thick*
½   *cup beef broth*
2   *tablespoons sugar*
    Generous pinch of saffron

Combine apple slices and broth in a saucepan and cover. Bring to a boil, reduce heat, and simmer for about ½ hour.

Mash apples well, stir in sugar and saffron, and simmer, covered, for 5 to 10 minutes more, stirring occasionally to keep sauce from scorching.

Serve hot with rich-flavored meats.

YIELDS ABOUT 1½ CUPS

SAUCE FOR FISH

⅓   *cup slivered blanched almonds*
1   *cup water*
3   *large apples, quartered*
½   *teaspoon fennel seeds*
2   *tablespoons sugar*

Put almonds and ½ cup water in a blender. Whirl for a minute or so until all the almonds have turned to a fine slush. Add the remaining ½ cup water and refrigerate for 2 hours.

Strain almond-water mixture through a very fine strainer or through a strainer lined with cheesecloth. Measure ½ cup of the resulting almond milk and combine with apple quarters in a saucepan.

Crush fennel seeds slightly using a mortar and pestle or crush with the back of a spoon on a hard surface. Add to the apples. Cover saucepan and bring to a boil over high heat, then reduce heat and simmer until apples are tender, about 35 to 45 minutes.

Force apples through a strainer, using the back of a spoon. Return to saucepan, stir in sugar, and cook over low heat to dissolve sugar and to bring the sauce back to serving temperature.

This is an unusual, tart-sweet sauce that goes well with poached freshwater fish.

YIELDS ABOUT 1½ CUPS

10

Beverages

APPLE TODDY
HOT MULLED CIDER
COLD MULLED CIDER
SYLLABUB
APPLE BLOSSOM CAUDLE
STONE FENCE PUNCH
APPLE WATER
APPLE LEMONADE
WASSAIL

Apple Toddy

*Boil a large juicy pippin in a quart of water, and when it
has broken to pieces strain off the water. While it is still boiling-
hot, add a glass of fine old whiskey, a little lemon-juice, and
sweeten to taste.*

Take hot at bed-time for influenza.

From *Common Sense in the Household*,
by Marion Harland, New York, 1884

1   *large apple*
1   *quart water*
½   *cup whiskey*
    Juice of 2 lemons
½   *cup sugar*

Quarter the apple without peeling or coring. Boil in water in a saucepan for about 45 minutes, strain, and add whiskey, lemon juice, and sugar to the apple water. Serve hot.

You don't have to have flu to enjoy this drink, but if you do, an Apple Toddy will take a bit of the edge off your suffering.

YIELDS APPROXIMATELY 1 QUART

Hot Mulled Cider

1   *quart sweet cider*
⅓   *cup firmly packed brown sugar*
1   *cinnamon stick*
8   *whole cloves*
¼   *teaspoon grated nutmeg*
¼   *teaspoon ground ginger*
1   *large lemon, sliced thin*
    Whole cloves for garnish

In a saucepan, combine cider, brown sugar, cinnamon stick, 8 whole cloves, nutmeg, and ginger.

Slice the lemon thin, pick out any seeds, and reserve 6 slices. Add the balance to the cider mixture and cover.

Put cider over heat and stir until sugar is dissolved. Bring to a boil, reduce heat, and simmer, covered, for 10 minutes.

Stud the reserved lemon slices with 1 or 2 whole cloves each. Strain cider into serving cups and garnish with clove-studded lemon slices.

YIELDS APPROXIMATELY 1 QUART

Cold Mulled Cider

To one quart of cider take three eggs. Beat them light and add sugar according to the acidity of the cider. When light, pour the boiling cider on, stirring briskly. Put back on the fire and stir till it fairly boils. Then pour it off.

From *Housekeeping in Old Virginia*,
edited by Marion Cabell Tyree,
Louisville, Kentucky, 1870

1   *quart sweet cider*
3   *eggs*
½   *cup sugar*

Put cider in a saucepan over moderate heat and bring to a boil.

Beat eggs in a mixing bowl until light and lemon-colored. Beat in the sugar.

Beat in a small quantity of the hot cider, then gradually beat in the rest. Return mixture to saucepan and place over moderate heat. Bring once more to a boil, then pour into serving cups. Serve hot.

SERVES 6

Syllabub

Fill your Sillabub pot with Syder (for that is the best for a Sillabub) and good store of Sugar and a little Nutmeg, stir it well together, put in as much thick Cream by two or three spoonful at a time as hard as you can, as though you milke it in, then stir it together exceedingly softly once about, and let it stand two houres at least ere it is eaten for the standing makes the Curd.

From *The Compleat Cook*,
by W. M.,
London, 1655

Syllabub had the consistency of soft, creamy yogurt and was eaten with a spoon as a dessert. When W.M. remarked that you should put in the cream "as hard as you can, as though you milke it in," he was referring to one of the commonest ways of making the dish in days gone by: The cook would take the syllabub pot out to the barn and milk a cow directly into the cider.

The success of this recipe depends on the formation of a curd, which is not the easiest thing to accomplish with modern pasteurized milk products. There is a way around this problem, however, and that is by the introduction of an ingredient that is rich with the bacteria that curdle milk: yogurt.

> 1   *cup sweet cider*
> ¼   *cup granulated sugar (optional)*
> 1   *pint heavy cream*
> 2   *tablespoons plain yogurt*
>   *Grated nutmeg to taste*

Heat cider in a saucepan until it is lukewarm. Stir in sugar, if used. (For many people, the dish will be sweet enough without sugar.)

Pour cider into a bowl and add the heavy cream. Stir the yogurt until it is soft, then add it to the cider and cream. Mix all together well. Cover bowl loosely and put in a warm place overnight.

The next day, refrigerate mixture for several hours, until curdled cream becomes quite firm.

Carefully skim the cream from the liquid beneath and spoon into demitasse or similar small serving cups. Perform these two operations as gently as possible, so as not to break up and liquify the curd. Top each serving with a little grated nutmeg.

SERVES 6 TO 8

Apple Blossom Caudle

*Tak crommys of wyte bred and the flowris of the swete
Appyltre and zolkys of Eggys and bray hem togedere in a morter
and temper yt up wyth wyte wyn and mak yt to sethe and wan yt
is thykke do thereto god spicis of gyngener, galyngale, canel and
clowys gelofre and serve yt forth.*

From *The Forme of Cury* (circa 1390),
edited by Samuel Pegge,
London, 1780

This recipe, translated from its original Middle English into
modern English, might sound something like this:

*Take the crumbs of white bread and the flowers of the sweet
apple tree and the yolks of eggs and pound them together in a
mortar. Then add white wine and put it on to cook. And when it
is thick, add good spices, such as ginger, galingale, cinnamon,
and cloves. Then serve it forth.*

Galingale is a mild, gingerlike spice that is not readily available
in the United States, so it has been omitted in the recipe that follows.
For some reason, either Pegge or the original compilers of the manuscript (the chefs of Richard II) identified this dish as a type of fritter.
However, there is no mention of frying in this recipe, so it isn't really
a fritter. Looking at the ingredients and the method of cooking, one
would have to call it a caudle, a hot liquor usually thickened with bread
crumbs and eggs. Caudles are classified as hot beverages, but often
they were thick enough to eat with a spoon, which is the case with the
recipe below.

1   *cup fine bread crumbs*
½   *cup chopped apple blossoms*
2   *egg yolks, well beaten*
2   *cups sweet white wine*
¼   *teaspoon ground ginger*
¼   *teaspoon ground cinnamon*
     Pinch of ground cloves

Combine all ingredients in a saucepan. Place over low heat and cook, stirring constantly, until smooth and thickened, about 10 minutes. Serve hot.

SERVES 2

Stone Fence Punch

Stone Fence Punch was a favorite with farm workers at the turn of the nineteenth century. It was concocted of sweet cider and rum or applejack. It is a rather potent drink, as the quantity of applejack indicates.

1   *quart applejack or dark rum*
2   *quarts sweet cider*
1   *orange, sliced*
10 to 15   *whole cloves*

Combine the applejack and cider in a punch bowl with a large chunk of ice or a molded ice ring. Garnish with orange slices, each studded with a few cloves.

For an individual serving, put 2 jiggers of applejack or rum into a highball glass. Add 2 ice cubes and fill the glass with sweet cider. A twist of lemon or orange peel adds flavor and color.

YIELDS 3 QUARTS

IN THE APPLE ORCHARD
An engraving from
Peterson's Magazine,
September 1872.

Apple Water

This is given as sustenance when the stomach is too weak to bear broth, &c. It may be made thus,—Pour boiling water on roasted apples; let them stand three hours, then strain and sweeten lightly;—Or it may be made thus,—Peel and slice tart apples, add some sugar and lemon-peel; then pour some boiling water over the whole, and let it stand covered by the fire, more than an hour.

From *The American Frugal Housewife*,
by Mrs. (Lydia Maria) Child,
Boston, 1833

This recipe was intended as a sort of broth for the sick, but it can make a cooling summer beverage.

4   *tart apples, peeled, cored, and thinly sliced*
¼   *cup granulated sugar*
      Peel of ¼ lemon
4   *cups boiling water*

Place apples, sugar, and lemon peel into a crock or pot with a close-fitting cover. Pour the boiling water over the apples, cover, and set aside to cool.

When it has reached room temperature, strain the liquid into glasses, add ice, and serve.

YIELDS APPROXIMATELY 1 QUART

Apple Lemonade

6   *apples, with stems and blossom ends removed, chopped (do not peel or core)*
1½ *cups water*
2   *cups sugar*
3   *lemons*
    Cold water

Combine apples with 1½ cups water in a covered saucepan and bring to a boil. Reduce heat and simmer until apples are tender, about 20 minutes.

Strain apples through a colander lined with two thicknesses of cheesecloth, reserving apple juice. Discard apple pulp. Measure out 2 cups of apple juice, adding a little water if necessary to make the 2 cups.

Put juice in a saucepan, add sugar, and place over low heat. Stir until sugar is dissolved. Let cool.

Divide the apple syrup equally into 6 glasses. Squeeze the juice of ½ lemon into each glass. Add ice cubes and fill glasses with cold water. Stir.

YIELDS 6 10-OUNCE GLASSES

Wassail

This recipe for the traditional Christmas drink of Britain is compounded of ale and sack (dry sherry) and is much closer to the early recipes and far less potent than most other modern versions, which are concocted of sherry and some kind of hard liquor. The recipe below will provide 24 servings (3 quarts). If the party is small, it is best to halve the recipe, because the mixture tends to separate after half an hour or so. The baked Lady apples used for garnish can be eaten and are delicious after they have soaked in the brew.

12	*Lady apples (or other very small apples)*
¼	*teaspoon whole cardamon seeds crushed*
6	*whole allspice, crushed*
2	*cinnamon sticks, broken into bits*
1	*tablespoon chopped and peeled fresh gingerroot*
10	*whole cloves*
1	*teaspoon grated nutmeg*
2	*quarts ale*
1	*fifth (⅘ quart) dry sherry*
2	*cups sugar*
6	*eggs, separated*

Preheat oven to 400 degrees F.

Bake apples whole in a baking dish at 400 degrees F. until tender, about 30 to 40 minutes. Set apples aside.

Crush cardamom seeds slightly with a mortar and pestle or with the back of a spoon on a hard surface. Make a bouquet garni of cardamom seeds, allspice, cinnamon, gingerroot, cloves, and nutmeg and tie with a string.

Pour 1 quart of the ale into a 4-quart or larger pot and add the spice bag. Place over heat and bring to a boil. Reduce heat and simmer for 10 minutes.

Remove spice bag and pour in remaining 1 quart ale. Add sherry and sugar, and simmer for 20 minutes more, stirring at first to dissolve sugar.

Warm punchbowl by filling it with hot water.

Beat egg whites in a mixing bowl until they are stiff enough to stand in peaks. In a separate mixing bowl, beat egg yolks until they are lemon colored, then fold them into the beaten egg whites.

After ale and sherry mixture has simmered 20 minutes, remove from heat and beat in egg mixture gradually.

Drain and dry punch bowl and fill with hot wassail. Float baked apples in the brew and serve immediately.

YIELDS 3 QUARTS

11

Jellies, Jams, and Preserves

CANNED APPLES

CANNED BAKED APPLES

CANNED APPLES FOR PIES

CANNED APPLESAUCE

PRESERVED APPLES

PICKLED APPLES

BOILED CIDER APPLESAUCE (APPLE BUTTER)

GINGER AND APPLE JAM

APPLE JELLY

MINT APPLE JELLY

METHODS OF PRESERVATION

As mentioned in Chapter 2, an apple is still "alive" after it has been taken from the tree. It continues to ripen and finally collapses in decay from processes going on inside the fruit itself. The goal of canning, drying, even of refrigerating apples is to slow or halt this inevitable march to destruction.

There are several ways of keeping or preserving apples for future use, as is the case with most foods. If the apple is nowhere near ripe, it can be kept in a cool place, which slows the process of ripening and deterioration. If the apple is ripe already, it can be refrigerated, which slows the process even more, or it can be frozen. Freezing quite literally kills the apple, stopping its life processes dead, and if the cold is intense enough, the fruit should last forever. (The home freezer is not quite cold enough to ensure apple immortality.)

Actually, freezing accomplishes two goals at once: It calls off the apple's self-appointed destruction, and it protects the fruit from the ravages of molds and bacteria that take over where the apple leaves off in the natural process of decay. Every other method of preservation has these same two goals, though the means differ.

When an apple is sliced and dried, the fruit's tissues become so dehydrated that life can no longer go on. Furthermore, the now leathery slices are practically impervious to the attacks of mold and bacteria as long as the slices are kept dry. When the apples are preserved in sugar, as in a jelly or a jam, the cooking halts the apple's natural decline, while the sugar protects the fruit from outside sources of decay. Pickling the apple amounts to the same thing, while in canning the fruit is cooked, sterilized, then kept sterile. The sterilization kills the decay-causing organisms; the closed container keeps more such organisms from getting to the apples.

There is nothing mysterious about any of these methods of preservation, and if you follow the directions carefully, success is almost assured. One of the chief reasons for the apple's continued popularity over the years has been the ease with which it can be kept and preserved.

UNREFRIGERATED STORAGE

If you grow your own apples, unrefrigerated storage is a major consideration. Choose the coolest storage place you can find. The general rule for storing fresh apples is: the cooler the better, all the way down to just above the freezing point of water (32 degrees F.). The cooler the storage place the longer the apples will keep. Apples of the so-called winter varieties, such as the Baldwin or the Winesap, will keep for months in a cool cellar. (For information on the keeping qualities of other varieties, see Chapter 2.)

The spot chosen for storage should be dry but not arid. Apples keep best when the air surrounding them is humid, but they should never get so damp that moisture collects on their skins. Too much moisture provides the proper growing conditions for molds and bacteria that cause decay.

The best containers for storing apples in a cellar are small wooden baskets with spaces between the slats. The spaces allow air to circulate between the apples, which prevents a buildup of moisture in the container. Small baskets are best simply because they are easier to handle than large baskets. There is no need to put wooden covers on the baskets unless you intend to stack one basket atop another. Pieces of cloth thrown over the tops of open baskets will keep the fruit from getting dusty and will help retain a healthy amount of moisture.

The fruit chosen for storage should be free of bruises and cuts. Damaged fruit decays more quickly, and the moisture and bacteria associated with such decay can damage the rest of the apples. There is an old and true adage: "One rotten apple can spoil the barrel."

REFRIGERATED STORAGE

If you grow one of the so-called summer varieties of apples—trees whose fruit ripen from late July into early September—you must consider refrigerated storage, unless you intend to consume or can the entire crop within a week or two. Summer varieties, such as Lodi and Yellow Transparent, are poor keepers, but their life expectancy as fresh fruits can be extended by several weeks if the apples are refrigerated immediately after picking. Winter varieties will keep months longer if they are chilled.

A refrigerator in the cellar is perfect for storing apples. Turn it to its coldest setting and jam it full of apples packed in plastic bags. If the bags aren't perforated, punch some holes in them yourself. If the bags are airtight, they will collect moisture, which is bad for the fruit.

Be careful, too, that you don't pile the apples so close to the freezing coils of the refrigerator that some of the fruits get frosted. You can freeze apples, of course, but this is not the way to do it. Furthermore, once an apple has been frozen, it loses its natural crispness and can no longer be used for eating out of hand or in salads.

Finally, remember that when apples have been stored half ripe in either cellar or refrigerator, it is quite likely that they will not be fully ripe when you take them up for use. This is not particularly important if you intend to cook the apples, but if you want them for dessert or salad use, take them upstairs two or three days ahead of time and keep them at room temperature, so that they will be ripe and mellow when wanted.

FREEZING

Apples, peeled and cored, can be frozen whole, in halves, or in quarters, but the usual way is to freeze slices of the size that go into apple pies. There are two ways of preparing the slices for freezing. Either method protects fruit quality and prevents discoloration during freezing and thawing.

Method 1: Peel, core, and slice the apples, then plunge the slices into boiling water. Cook for 1½ to 2 minutes, then remove and drain.

Method 2: Peel, core, and slice the apples, then dip for 1 minute in a mixture of 3 tablespoons of lemon juice to 1 gallon (4 quarts) of water. Rinse the slices in cold water and drain.

The slices can then be packed either dry or with sugar (1 cup sugar for each 3 to 5 cups of slices, depending on taste) in plastic-lined paper containers, in plastic containers, or in straight-sided, shoulderless, canning jars. Pack each container nearly to the top, leaving a little room for expansion, and freeze immediately.

Place the containers as close as possible to the walls of the freezer. This will freeze the apples in the shortest amount of time and do the least violence to the texture of the fruit. When apples are frozen slowly, ice crystals form in the tissues, breaking up the internal structure of the fruit. The result is mush when the slices thaw.

Applesauce can be frozen with no further preparation than packing in containers, leaving a little room for expansion. This is a nice way to preserve a few pounds of windfall apples on a hot summer day, when you don't want to heat up the kitchen any more than is absolutely necessary.

Whether you are freezing slices or sauce, don't go loading up the freezer with pound after pound of unfrozen fruit. If you overload the equipment, freezing will be slow and fruit quality will suffer. A rule of thumb is that during any 24-hour period you should not freeze more than 3 pounds of food per cubic foot of freezer capacity. The typical free-standing home freezer has a capacity of 15 to 20 cubic feet. A quart container of dry apple slices weighs about 1 pound, while the same container weighs 1½ pounds when filled with sugared slices and 2 pounds when filled with applesauce. The freezer itself must be capable of maintaining a temperature of 0 degrees F. or lower at all times.

Although freezing should be fast, thawing should be as slow as possible to maintain quality. Thaw slices and sauce in their original containers before using, preferably overnight in the refrigerator. Thawed frozen apples can be used in any cooked or baked dish calling for fresh apples. Just remember that if you packed the apple slices with sugar, you must reduce the sugar in the recipe accordingly. In pies, for example, most people would not want more sugar than is already present in the frozen apples.

DRYING

Drying is one of the oldest ways known to man of preserving food. Meats, vegetables, and fruits were dried for winter use long before the people who dried them learned to write and keep records. Apples will keep for weeks and months if stored in a cool place, but they will survive for months or years if they are dried.

Drying an apple is simplicity itself. The fruit usually is cut into pieces and cored. Whether it is peeled is a matter of choice, usually depending on the toughness of the skin. The apple slices are then put in a warm place where they will dry before they can go moldy. In days gone by, apples were dried by placing them in the sun or by hanging them near a chimney or under the roof beams of a shed or house. This latter method is still practical today. If you have an attic that is as hot as blazes in summer, you have an ideal spot for drying apples.

Peel and core the fruit and cut into eighths. Thread a large-eyed needle with light cotton twine and pass it through the middle of each slice. String the slices like beads, 50 or more slices to a string, then tie up the strings like so many clotheslines under the roof of the house.

The attic should be hot but well ventilated, with screens over the windows to keep out bugs. Depending on the weather, the apples will be dry in a matter of a week or two, but as long as the weather stays hot (and the apples dry), it doesn't hurt to let them hang. When you think the slices are perfectly dry (you can cut a few slices in half to be sure), pack them in airtight containers and store in a cool, dry place.

To use dried apples, cover them with fresh water and soak overnight, then add them to cooked or baked recipes as you would fresh apples. For maximum flavor, use the water in which the apples were soaked whenever the recipe specifies water. One pound of dried apples (before soaking) is the equivalent of 3½ to 4 pounds of fresh apples.

JELLIES, JAMS, MARMALADES, AND PRESERVES

Jams, jellies, marmalades, and preserves have one thing in common: They all depend on sugar for their keeping qualities. However, sugar isn't the only key to a good jelly. The fruit must contain the right amount of pectin, or the jelly or jam simply won't gel. The fruit also must contain a certain amount of acid, or the end product will be soft and runny. Apples, especially when they are slightly underripe, contain sufficient quantities of pectin and acid to make a good apple jelly or preserve with only three ingredients: apples (obviously), sugar, and water. Other ingredients are added only for flavor. It's small wonder, then, that apples figure in some of the earliest preserving recipes.

CANNING AND PRESERVING

Preserving whole apples and apple pieces has been going on in the home since the mid-seventeenth century, at least. Canning was a nineteenth-century development and was not commonly practiced at home until after the Civil War, when glass canning jars became popular. There are certain differences between the two processes, and Fannie Farmer summed them up quite neatly in *The Boston Cooking-School Cook Book:*

> Preserving fruit is cooking it with from three-fourths to its whole weight in sugar. By doing so, much of the natural flavor of the fruit is destroyed; therefore canning is usually preferred to preserving.
> Canning fruit is preserving sterilized fruit in sterilized air-tight jars, the sugar being added to give sweetness. Fruits may be canned without sugar if perfectly sterilized, that is, freed from all germ life.

We may disagree a little with Miss Farmer on whether canning is "usually preferred" over preserving, since in this world of varying flavors there should be room for all. However, there is no question that canned apples taste more like cooked fresh apples than do apple preserves. Neither, of course, really tastes like the fresh, uncooked fruit.

The canning process has changed little in a century. If the raised letters on old jars are accurate, John Mason of New York patented his namesake in 1858. His patent was for a screwtop glass jar with a glass-lined metal cap. Jars almost identical to the original are in use today.

Canning recipes also have changed little, so there would be no great interest in quoting old recipes in that department.

Canned Apples

There really is no specific recipe for canned apples, but there is a definite plan of action that must be followed.

First, get a water-bath canning pot (canner), which is a large pot equipped with a rack in the bottom and a tight-fitting lid. (Apples need not be canned in a pressurized pot.)

Next, determine how many jars can be processed in your canner at once. The jars cannot be jammed in. There must be space between the jars and the outside of the pot, and there must be space between the jars. Do your canning in one canner-load batch at a time, because success depends on getting hot fruit and syrup into hot jars and processing those jars in the canner as quickly as possible.

For each quart jar, you will need 2 to 2½ pounds of fresh, whole apples. Choose fruit that is sound, firm, and ripe. Wash the apples, then peel and core them, cutting them into halves, quarters, or slices as desired. To keep apples from turning brown after peeling, drop them into a mixture of 2 tablespoons cider vinegar, 2 tablespoons salt, and 1 gallon (4 quarts) cold water. Peel only enough fruit for one batch, because the fruit will get mushy if left in the solution too long.

For each quart jar you will also need 1 to 1½ cups syrup, made by combining 1 cup sugar with 2 cups water. Combine the ingredients in a large saucepan and cook until sugar is dissolved. Always prepare a little more syrup than you think you will need. Jars must be filled with syrup to ½ inch of the top, and it is better to have too much syrup than too little. It is a wise policy to have extra hot syrup in a small saucepan to top off jars. Keep this syrup hot, but don't let it boil down.

When syrup is ready and apples have been prepared, remove the fruit from the salt and vinegar water. Rinse with fresh water and drain. Put the apples into the syrup, bring to a boil, and boil for 5 minutes. Pack the apples into hot, sterilized canning jars (prepared according to manufacturer's instructions) and fill with syrup to ½ inch of the top. Run a knife blade between the glass and the fruit to get out any air bubbles. Be careful not to scratch the inside of the jar with the blade, or the jars might break during processing. Wipe off the mouth and neck of each jar and cap loosely.

Have the canning pot about a third filled with boiling water and have a kettle filled with more boiling water on hand. Put the jars on the rack in the canner, leaving space all around. Fill the canner with more boiling water to about 1 inch from the bases of the jar caps. (You cannot process jars of widely different heights, such as quarts and pints, in the same batch.) Cover canner and bring to a full boil. Start timing at this point only, and process jars for 20 minutes.

After processing, remove jars to a rack, arranging them so there is space between each jar. As soon as each jar is taken from the hot water, tighten the cap fully, sealing the jar. After jars have cooled, test seals according to manufacturer's instructions for the particular jar used, then store in a cool, dark place. Canned apples, jellies, and preserves will keep their color best when stored in a dark place at a temperature between 45 and 60 degrees F.

Some cautions: Have everything you will need—spoons, pot holders, a long-bladed knife, jars and lids, a clean rag for wiping jar necks—prepared and on hand before you begin cooking a batch of fruit. Syrup for several batches can be prepared in advance, then heated to boiling as needed. Be sure the syrup is at a full boil before putting in the apples. Work as quickly as possible once you begin the cooking-to-processing sequence. Once the jars are in the canner, you can begin preparing the apples for the next batch of jars. When you reach the end of your apple and jar supply, you may not have a full batch of jars for the canner. This doesn't matter. You can overfill a canner with jars, but you can't underfill it.

Canned apples can be eaten for dessert "as is," or they can be substituted in many recipes based on stewed apples. Just omit any steps designed to cook the apples, though they can be heated, and omit any sugar in the recipe that is added directly to the apples. These apples can be used for pies, but many people find them too sweet. However, apples can be canned specifically for pies (see Canned Apples for Pies, page 222).

Canned Baked Apples

Prepare a sugar-and-water syrup, using proportions of 1 cup sugar to 2 cups water. Wash and core the apples and put them into a baking dish. Pour syrup into the pan to a depth of about ¼ inch—a little more won't hurt. For each 10 apples, combine ¼ cup sugar and ½ teaspoon ground cinnamon (or 1 teaspoon sugar and a generous pinch of ground cinnamon for each apple). Sprinkle this mixture over the apples.

PARING, CORING, AND SLICING MACHINE

This hand-powered, cast-iron wonder represented the last stage in the development of a device once found in every farmhouse kitchen. Wooden versions were manufactured in the Colonies as early as 1750, but these earlier models were capable only of paring the fruit. From S. E. Todd's *The Apple Culturist*, New York, 1871. (*Courtesy the Library of the New York Botanical Garden*)

Bake at 400 degrees F. until apples are half done, about 35 to 40 minutes.

Prepare canning jars according to manufacturer's instructions, and just before removing apples from oven, put more syrup on the heat to boil. Pack hot apples into hot, sterilized jars. Pour in boiling hot syrup—from baking pan and extra syrup as needed—to fill each jar to within ½ inch of top. Cap jars loosely, and process in boiling water in a canner for 20 minutes (see Canned Apples, page 219).

Canned Apples for Pies

Follow the instructions for Canned Apples, page 219, but substitute a syrup made by combining 1 cup of sugar with 4 to 5 cups of water. Apples canned in a syrup this light tend to be a little less firm than those canned in a heavier syrup. However, they are suitable for use in pies (see Pie with Canned Apples page 236).

Canned Applesauce

Wash apples. Remove stems and blossom ends.

Cut apples into eighths and put into a large pot with about 1 cup of water to keep them from scorching. Cook apples until tender, about 35 to 40 minutes.

Force apples through a coarse strainer with the back of a large spoon, or run them through a food mill to remove skins and cores.

Sweeten sauce to taste, but don't add spices, because they will darken the sauce. Put sauce back in cooking pot and reheat to boiling.

Pour at once into hot, sterilized canning jars (prepared according to manufacturer's instructions), filling to within ⅛ inch of top. Cap jars loosely, and process in boiling water in a canner for 20 minutes (see Canned Apples, page 219).

Preserved Apples

Weigh equal quantities of Newtown pippins, and the best of sugar; allow one sliced lemon for every pound. Make a syrup, and then put in the apples. Boil them until they are tender; then lay them into the jars and boil the syrup until it will become a jelly. No other apple can be preserved without breaking. This keeps its shape, and is very beautiful. Quarter the apples, or take out the core and leave them whole, as you prefer. Other sour hard apples are very good preserved, but none keep as well, or are as handsome as the Newtown pippins.

From *The Young Housekeeper's Friend*,
by Mrs. (Mary Hooker) Cornelius,
Boston, 1859

Most people would find the amount of lemon specified in this recipe overpowering, so in the updated version the amount has been reduced. A few slices of peeled fresh gingerroot can be cooked with the apples, if desired.

      2   *lemons*
          Gingerroot to taste (optional)
 3⅓   *cups water*
      5   *pounds sugar*
      5   *pounds apples, peeled, cored, and quartered*

Slice lemons into rounds about ⅛-inch thick and pick out seeds. Set aside.

Peel and thinly slice gingerroot, if used.

Combine water, sugar, lemon rounds, and sliced gingerroot in a large saucepan. Place over high heat, bring to a boil, and boil for 3 minutes.

Reduce heat, add apples, and cook gently until apples are tender and transparent. Remove hot apples and pack into approximately 5 hot, sterile 1-pint canning jars (prepared according to manufacturer's instructions). Put a slice or two of lemon into each jar (near the glass, so it will show) "for pretty."

Heat the syrup to a rolling boil, pour over apples, and fill jars to within ⅛ inch of top. Seal immediately. Cool jars, well separated from one another, on a rack.

YIELDS APPROXIMATELY 5 1-PINT JARS

Pickled Apples

Best vinegar, 1 gal.; sugar, 4 lbs.; apples, all it will cover handsomely; cinnamon and cloves, ground, of each 1 tablespoon.

Pare and core the apples, tying up the cinnamon and cloves in a cloth and cooking until done, only. Keep in jars. They will be nicer than preserves, and more healthy, and keep a long time; not being too sour, nor too sweet, but an agreeable mixture of the two.

From *Dr. Chase's Recipes*,
by A. W. Chase,
Ann Arbor, Michigan, 1874

When preparing whole apples for pickling, it is best to core them first, then to peel them, cutting away the dried blossom remnants at the base of each fruit. Apples can be kept from turning brown during preparation by dropping them into a mixture of 1 gallon of cold water, 2 tablespoons cider vinegar, and 2 tablespoons salt.

4 *cups cider vinegar*
2 *cups water*
6 *cups sugar*
2 *tablespoons whole cloves*
4 *cinnamon sticks, broken into bits*
8 *pounds small apples, cored and peeled*

In a large pot, combine cider vinegar, water, and sugar. Combine the cloves and cinnamon sticks in a bouquet garni and drop into the liquid. Place pot over high heat and boil until sugar is dissolved. Reduce heat, add apples, and simmer until fruit is just tender, about 20 to 30 minutes. Remove pot from heat, cover, and let stand for 12 to 18 hours in a cool place.

Remove apples from syrup and pack into approximately 7 hot, sterilized 1-pint canning jars (prepared according to manufacturer's instructions).

Remove spice bag and reheat syrup to a rolling boil. Pour the syrup, while boiling hot, over apples in jars, filling to within ⅛ inch of top. Cap jars loosely and process at once in boiling water in a canner for 10 minutes (see Canned Apples, page 219).

YIELDS APPROXIMATELY 7 1-PINT JARS

Boiled Cider Applesauce (Apple Butter)

Take apples, sweet and sour together, that will not keep long, and pare a large quantity. When finished, wash and put them into a large brass kettle, in which you have turned down an old dish or large plate, that will nearly cover the bottom; this is to prevent the apple from burning. After you have put in all the apples, pour in a quart of cider (boiled as directed in the receipt for boiled cider) to every pailful of apples. After it has boiled an hour or two, add molasses in the proportion of two quarts to every four pails of apples. If you have refuse quinces, a peck of them gives a fine flavor to a large kettle of apple-sauce. The best way to boil apple-sauce is to put the kettle over the fire at night, and let the apple become partly done before bed-time. When you leave it for the night, see that the fire lies in such a way, that all parts of the apple boil equally, and that no brands can fall.* Burn charcoal or peat if you have it, as either of these will make a steady fire, and may be left without danger from snapping. The

---

* As the open fire-place is now seldom in use, these directions will not often be apropos. But where a range or coal stove is used, a large kettle of apple-sauce can, with care, be done well, on the top with the cover under it.

chief things to be observed, are, that there is not too much fire, that it lies safely, and that it will afford a moderate heat several hours. In the morning the apple-sauce will be of a fine red color, and must then be put away in firkins or stone jars. Never use potter's ware *for this purpose.*

From *The Young Housekeeper's Friend,*
by Mrs. (Mary Hooker) Cornelius,
Boston, 1859

Benjamin Thompson, a Massachusetts-born Tory who fled to England in 1776, invented a closed-top cooking range in the early 1800s. But it wasn't until the late 1850s that coal ranges became standard equipment in middle-class urban homes in America. After 1880, as coal gas was piped into an increasing number of homes, coal ranges were slowly replaced by gas ranges, a process that took more than half a century.

The molasses of Mrs. Cornelius's recipe is replaced by brown sugar in the modernized version below. The spicing is optional and can be varied or eliminated to suit one's taste.

1   gallon sweet cider
20   pounds apples (just under ½ bushel)
2   cups water
3   cups firmly packed dark brown sugar
5   tablespoons ground cinnamon
2   tablespoons ground cloves
1   tablespoon ground allspice

Put cider in a large pot and boil gently until volume has been reduced to about 2 quarts.

Peel, quarter, and core apples, then cut them into slices. Combine apples with 2 cups of water in a separate covered pot, bring to a boil, then reduce heat and simmer until apples are tender, about ½ hour. Mash apples, or run through a food mill if an extremely smooth butter is desired.

By the last quarter of the nineteenth century the coal range had superseded the fireplace in American cookery. However, the change did not represent a step forward in terms of convenience, since the coal range was every bit as difficult to use. Fireplace cookery went out of style principally because firewood had become prohibitively expensive after the destruction of the great hardwood forests of the East. From J. A. and R. A. Reid's *Picturesque Narragansett*, Providence, Rhode Island, 1888.

Preheat oven to 325 degrees F.

Combine apples and boiled cider in a large, flat roasting pan. Place in a preheated oven and roast, stirring occasionally to prevent scorching, until apples are reduced by half, about 1½ hours.

Add brown sugar and stir well until dissolved. Cook for 1 hour more, stirring every 5 to 10 minutes to prevent scorching.

Stir in ground cinnamon, ground cloves, and ground allspice and cook until done, about ½ hour more. Test apple butter by putting a spoonful on a plate. If no ring of liquid forms around the edges of the butter on the surface of the plate, apple butter is done. Pour immediately into approximately 6 hot, sterilized 1-pint canning jars (prepared according to manufacturer's instruction), filling each to within ⅛ inch of top. Seal at once. Cool jars, well separated from one another, on a rack.

YIELDS APPROXIMATELY 6 1-PINT JARS

Ginger and Apple Jam

Weigh equal quantities of brown sugar and good sour apples. Pare and core them, and chop them fine. Make a syrup of the sugar, and clarify it very thoroughly; then add the apples, the grated peel of two or three lemons, and a few pieces of white ginger. Boil it till the apple looks clear and yellow. This resembles foreign sweetmeats. The ginger is essential to its peculiar excellence.

From *The Young Housekeeper's Friend,*
by Mrs. (Mary Hooker) Cornelius,
Boston, 1859

The recipe that follows puts more emphasis on the ginger and less on the lemon. If you are going to use a candy thermometer, check the boiling point of water for that day. The boiling point rises and falls with the barometer. If you don't have a candy thermometer, cook the jam until it has thickened somewhat, remembering that it will get even thicker as it cools. If fresh gingerroot is unavailable, substitute an equal quantity of minced candied ginger.

   1 *lemon*
1 1/2 *cups water*
   5 *cups firmly packed brown sugar*
   8 *cups peeled, cored, and chopped apples*
 1/2 *cup peeled and minced fresh gingerroot*

Quarter the lemon and pick out any seeds. Slice thin and set aside.
   Combine water and brown sugar in a large, heavy pot. Set candy thermometer in place. Cook over high heat until sugar is dissolved.
   Add chopped apples, lemon slices, and gingerroot to the syrup. Boil rapidly, stirring constantly, until temperature rises 9 degrees F. above the boiling point of water for that day, or until the mixture has

thickened. Remove from heat and alternately stir and skim the mixture for 5 minutes.

Ladle jam into approximately 6 hot, sterilized ½-pint canning jars (prepared according to manufacturer's instructions), filling each to within ⅛ inch of top. Seal immediately. Cool jars, well separated from one another, on a rack.

YIELDS APPROXIMATELY 6 ½-PINT JARS

Apple Jelly

To make Gelly of Pippins, take Pippins and pare them and quarter them and put as much water to them as will cover them, and let them boyl till all the vertue of the Pippins are out, then strain them, and take to a pint of that liquor a pound of sugar, and cut long threads of Orange peels, and boyl in it, then take a Lemon and pare and slice it very thin, and boyl it in your liquor, a little thin, take them out, and lay them in the bottome of your glass, pour it in the Limons in the glass. You must boyle the Oranges in two or three waters before you boyl it in the gelly.

From *The Queens Closet opened*
by W. M.,
London, 1655

3   pounds tart apples, about a quarter of them underripe
5   cups water
    Peel of 1 lemon, sliced very thin
    Peel of 1 orange, sliced very thin
    Juice of 1 lemon
3   cups sugar

Remove stems and blossom ends from the apples and cut apples into small pieces. Do not peel or core. The peels and pits contain pectin, which is essential to the proper setting of the jelly.

Place apples in a large covered pot, add 3 cups of water, and bring to a boil over a high heat. Keep the pot covered. When the water boils, reduce heat and simmer for about 25 minutes.

Mash the cooked apples and place in a jelly bag or in a colander lined with several thicknesses of cheesecloth. Allow to drip overnight, reserving the juice. Discard apple pulp.

The next day, put sliced lemon and orange peels in a saucepan with 2 cups of water, bring to a boil, and simmer for ½ hour to remove bitterness. Reserve peel and discard water. Set aside.

Measure out 4 cups of the apple juice into an 8-quart kettle. Add the lemon juice and the prepared peel. Then add the sugar and stir well until it is dissolved. Place kettle over high heat and bring to a rapid boil. The mixture must be boiled until it reaches the jelling point, which can be determined by any of the following three tests:

1. *Temperature Test*: Just before cooking, check the boiling point of water (it varies with daily atmospheric pressure) with a candy thermometer or deep-fat thermometer. When the fruit juice-sugar syrup reaches a point of 8 degrees F. above the water's boiling point for that day, the jelly is done.

2. *Spoon Test*: Take a cool metal spoon and dip it into the boiling syrup. Raise the spoon about a foot above the pot and turn it so that the syrup runs off the side of the spoon. If two drops form that flow together and fall from the side of the spoon in a single sheet, the jelly is ready.

3. *Saucer Test*: Put several saucers in the refrigerator about an hour before you start cooking the jelly. Place a spoonful of the boiling syrup on a chilled saucer and put it in the freezer for 2 to 3 minutes. If the jelly sets, it is ready. (Remove the kettle of syrup from the heat while making this test.)

When the jelly is done, remove it from the heat and skim off foam with a metal spoon. Quickly ladle jelly into approximately 7 hot, sterilized 6-ounce canning jars or glasses (prepared according to manufacturer's instructions), filling them to within ⅛ inch of top. Cover with a ⅛-inch-thick layer of melted paraffin. Seal immediately. Cool jars, well separated from one another, on a rack.

YIELDS APPROXIMATELY 7 6-OUNCE JARS

Mint Apple Jelly

Follow the directions for Apple Jelly (page 229), but eliminate the lemon and orange peel, the 2 cups of water used to cook the peel, and the steps involved in preparing the peel.

Prepare instead ¼ cup of fresh mint leaves by bruising them in a bowl with the back of a large spoon, then tie them in a bag made of two thicknesses of cheesecloth. Put the bag into the kettle with the apple juice, lemon juice, and sugar. Cook according to directions for Apple Jelly. Just before removing jelly from heat, stir in 4 to 5 drops of green food coloring, if desired. When jelly is done, remove mint bag, and proceed with canning as for Apple Jelly.

YIELDS APPROXIMATELY 7 6-OUNCE JARS

12
Pastry

PIE PASTRY

APPLE PIE

PIE WITH CANNED APPLES

APPLE TARTS

PUMPKIN AND APPLE PIE

APPLE AND RHUBARB PIE

APPLE CUSTARD PIE

TAFFETY-TART

CRANBERRY AND APPLE PIE

BEEF MINCEMEAT PIE

MEATLESS MINCE PIE

APPLE AND GREEN TOMATO PIE

CIDER CAKE

APPLE CAKE

APPLESAUCE CAKE

DRIED-APPLE CAKE

Pie Pastry

The following ingredients will yield enough dough for a 2-crust 8- or 9-inch pie or for 8 3-inch tarts. Halve the ingredients to make enough dough for a 1-crust pie.

2   *cups sifted flour*
¾   *teaspoon salt*
⅔   *cup shortening, chilled*
4 to 6   *tablespoons ice water*

Sift together flour and salt into a large mixing bowl.

Put chilled shortening atop this mixture and cut in with a pair of knives or a pastry blender until the largest piece of shortening is the size of a small pea.

Add ice water, a tablespoon at a time, tossing the mixture lightly with a pair of forks after each addition. Knead the moistened section of the dough to one side, so that each time water is added it can be sprinkled over a dry area. Add ice water until a small portion of the dough, squeezed lightly, forms a ball.

Gather dough together into a ball, flatten the ball slightly, and chill in the refrigerator for 10 minutes.

Divide dough into two slightly unequal portions. The smaller portion will be used for the bottom crust. Place smaller portion on a flat, lightly floured surface. Roll out with a lightly floured rolling pin to a thickness of ⅛ inch. Always roll from the center of the dough toward the edges. Do not roll back and forth; this will stretch and tear the dough. Roll out till the round of the dough is about 2 inches bigger than the pie pan. Fold the dough in half, then in half again.

For a pie, place the smaller portion of the dough in the pie pan so that the folded corner of the dough is at the center of the pan. Unfold the dough so that the pan is covered. If the pastry hasn't settled into the corners at the bottom of the pie pan, don't use force. Gently lift the pastry from the edges of the pan and set it down again, allowing the pastry to sink into the corners. Roll out the top crust and fold in half. Put filling into pie pan, then moisten the edges of the bottom crust slightly with water. Lay the folded top crust over half the filling, unfold, then press lightly around the rim of the pie pan to seal the two crusts together. Trim the dough with a sharp knife, leaving about 1 inch excess all around the pie. Roll up the excess crust so it looks as though a piece of rope had been laid around the rim of the pie pan. Flute the edge, if desired, with the fingers of one hand, placing index and middle fingers on one side of the rolled up crust and thumb on the

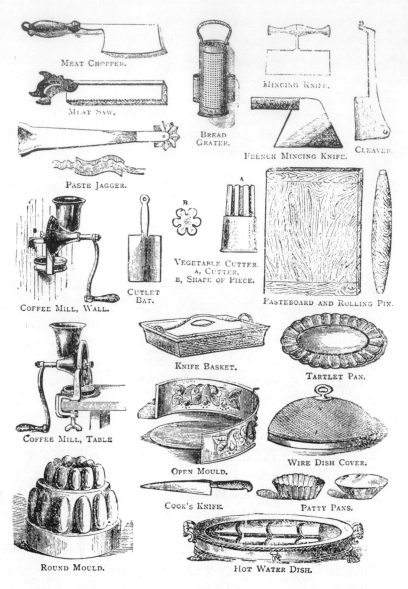

MEAT CHOPPER.

MEAT SAW.

BREAD GRATER.

MINCING KNIFE.

FRENCH MINCING KNIFE.

CLEAVER.

PASTE JAGGER.

COFFEE MILL, WALL.

CUTLET BAT.

VEGETABLE CUTTER.
A, CUTTER.
B, SHAPE OF PIECE.

PASTEBOARD AND ROLLING PIN.

KNIFE BASKET.

TARTLET PAN.

COFFEE MILL, TABLE

OPEN MOULD.

WIRE DISH COVER.

ROUND MOULD.

COOK'S KNIFE.

PATTY PANS.

HOT WATER DISH.

The kitchen utensils of the late nineteenth century should look familiar to most twentieth-century cooks and especially to practitioners of fine French cookery, where many of these devices are still in everyday use. From *Mrs. (Isabella Margaret) Beeton's Book of Household Management*, London, 1891.

other. Squeeze lightly, allowing the thumb to pass between the two fingers and shaping the dough into a little *V*. Repeat all around the edge of the pie. Slash crust in several places to let out steam. Bake as directed in individual recipe.

Apple Pie

(Pie Pastry for a two-crust pie page 232)

4 or 5   *{large} sour apples*
⅓   *cup sugar*
¼   *teaspoon grated nutmeg*
⅛   *teaspoon salt*
1   *teaspoon butter*
1   *teaspoon lemon juice*
    Few gratings lemon rind

Line pie plate with paste. Pare, core, and cut the apples into eighths, put row around plate one-half inch from edge, and work towards centre until plate is covered; then pile on remainder. Mix sugar, nutmeg, salt, lemon juice, and grated rind, and sprinkle over apples. Dot over with butter. Wet edges of under crust, cover with upper crust, and press edges together.

Bake forty to forty-five minutes in moderate oven. A very good pie may be made without butter, lemon juice and grated rind. Cinnamon may be substituted for nutmeg. Evaporated {dried} apples may be used in place of fresh fruit. If used, they should be soaked over night in cold water.

From *The Boston Cooking-School Cook Book*,
by Fannie Merritt Farmer,
Boston, 1896

This pie is perhaps a little light on the butter (1 to 2 tablespoons would be more usual) but otherwise is about as American as an apple pie can get. The recipe is self-explanatory and perfectly up to date, except in one area: Today, with the easily controlled temperatures of a gas or electric oven, standard procedure is to bake the pie in a preheated oven at 450 degrees F., then to reduce the heat to 350 degrees F. after 15 minutes and bake the pie for 35 to 40 minutes more.

YIELDS 1 9-INCH PIE

Pie with Canned Apples

Pie Pastry for a two-crust pie (see recipe on page 232)

Filling:
3 to 3½	cups Canned Apples for Pies (see page 222)
½	cup sugar
1	tablespoon cornstarch
1	teaspoon ground cinnamon
	Juice of ½ lemon
⅓	cup syrup from Canned Apples recipe
2	tablespoons butter

Preheat oven to 400 degrees F.

Line a 9-inch pie pan with approximately half the pastry dough. Put in a layer of canned apple slices.

In a mixing bowl, combine sugar, cornstarch, and cinnamon and mix well. Sprinkle about half this mixture over the layer of apples. Sprinkle on about half the lemon juice.

Fill the pie with the remaining apple slices, heaping them slightly in the center. Sprinkle on the rest of the sugar mixture and lemon juice. Pour in the syrup from the canned apples and dot with butter.

Put on the top portion of pastry dough. Seal, trim, and flute the edges, slashing the top in several places to let out steam. Bake in preheated oven until crust is nicely browned, about 35 minutes.

YIELDS 1 9-INCH PIE

Apple Tarts

To make all manner of Fruite Tarts you must boile your fruite, whether it be apple, cherie, peach, damson, peare, Mulberie, or codling, in faire water, and when they be boyled enough, put them in a bowle and bruse them with a ladle, and when they be colde, straine them, and put in red wine or claret wine, and so season it with sugar, sinamon and ginger.

From *The Good Husvvifes Ievvell*,
by Thomas Dawson, London, 1587

Filling:

6	*large apples*
½	*cup water*
1	*cup dry red wine*
1	*cup sugar*
2	*teaspoons ground cinnamon*
1	*teaspoon ground ginger*
2	*tablespoons butter (optional)*
2	*eggs, lightly beaten (optional)*

12 *unbaked tart shells or 1 9-inch pie shell (see Pie Pastry for a one-crust pie, page 232)*

To make filling, gently boil the apples (it is not necessary to pare, core, or slice them) in water in a covered saucepan for about ½ hour or until fruit is very tender. Drain off the juice. A potato masher will work fine for "bruising" (mashing) the fruit, which should then be worked through a strainer with the back of a spoon.

Preheat oven to 425 degrees F.

Flavor the fruit with the wine, sugar, cinnamon, and ginger. (Many would welcome the addition of 2 tablespoons of butter, and since this mixture has the consistency of applesauce, it can be tightened with 2 lightly beaten eggs.)

Pour the mixture into eight 3-inch tart shells and bake at 425 degrees F. for 10 minutes. Reduce heat to 350 degrees F., and bake for 30 minutes more.

YIELDS EIGHT 3-INCH TARTS
OR 1 9-INCH PIE

Pumpkin and Apple Pie

To make a Pumpion Pye, take about half a pound of Pumpion and slice it, a handfull of Tyme, a little Rosemary, Parsley and sweet Marjorum slipped off the stalks, and chop them small, then take Cynamon, Nutmeg, Pepper, and six Cloves, and beat them, take ten Egges and beat them, then mix them, and beat them altogether, and put in as much Sugar as you think fit, then fry them like a Froize, after it is fryed, let it stand till it be cold, then fill your Pye, take slices Apples thinne round wayes, and lay a row of the Froize, add a layer of Apples with Currants betwixt the layer while your Pye is fitted, and put in a good deal of sweet Butter before you close it, when the Pye is baked, take 6 yelks of Eggs, some white wine or Vergas and make a caudle of this, but not two thick, cut up the lid and put it in, stir them well together whilst the Eggs and Pumpions be not perceived and so serve it up.

From *The Compleat Cook*,
by W. M.,
London, 1655

The recipe that follows is a somewhat simplified adaptation of the original pie. In the seventeenth-century version, the pumpkin was sliced thin, sweetened and spiced, then fried with beaten eggs into a sort of omelet, called here a "froize." The fraise (to use a more modern spelling), was allowed to cool and then was sliced. This was combined with sliced apples and currants in a pie pan, along with a top crust. The recipe doesn't say so, but the rim of the pie pan would be floured to keep the crust from sticking. After the pie was baked, the top crust was carefully lifted, and a caudle (a sort of thin custard, in this case made with egg yolks and wine), was poured in. The crust was replaced, the pie baked a little longer, then served.

In this modern, two-crust version of the pie, the fraise has been replaced with sautéed pumpkin, to give the pie something of the fried flavor of the fraise. The caudle has been replaced by putting the eggs and wine into the pie shell at the outset, eliminating the need to lift and replace the top crust.

Filling:

3	cups cubed fresh pumpkin
3	tablespoons butter
1	teaspoon ground cinnamon
½	cup sauterne or similar sweet white wine
1	cup sugar
2	eggs, lightly beaten
1	teaspoon minced fresh parsley
½	teaspoon thyme
¼	teaspoon rosemary
¼	teaspoon marjoram
¼	teaspoon black pepper
¼	teaspoon salt
3 to 4	large apples, peeled, cored, and sliced thin
½	cup seedless raisins

Pie Pastry for a two-crust pie (see recipe on page 232)

Preheat oven to 450 degrees F.

Sauté pumpkin cubes in butter in a large skillet until pumpkin is tender enough to mash.

Mash pumpkin in a large mixing bowl. Add the cinnamon, wine, and sugar, and mix thoroughly. Beat in lightly beaten eggs. Add parsley, thyme, rosemary, marjoram, black pepper, and salt, and mix well. Add apple slices and raisins, and mix well.

Line a 9-inch pie pan with approximately half the pastry dough. Pour in the filling. Put on the top portion of pastry dough and seal and flute the edges, slashing the top in several places to allow steam to escape.

Bake at 450 degrees F. for 15 minutes. Reduce heat and bake at 350 degrees F. for 40 minutes more.

YIELDS 1 9-INCH PIE

Apple and Rhubarb Pie

Filling:

2 to 3	*large stalks rhubarb, cut into 1-inch-long pieces*
2	*large apples, peeled, cored, and sliced thick*
⅓	*cup sugar*
1	*tablespoon grated orange peel*
½	*teaspoon ground cloves*
1½	*tablespoons cornstarch*

Pie Pastry for a two-crust pie (see recipe on page 232)

Preheat over to 450 degrees F.

Combine all ingredients for filling in a large mixing bowl. Toss all together well.

Line a 9-inch pie pan with approximately half the pastry dough. Put in the filling. Cover with top portion of pastry dough. Seal, trim, and flute the edges of the crust, slashing the top in several places to allow steam to escape.

Bake at 450 degrees F. for 15 minutes, then reduce heat to 350 degrees F. and bake 40 to 45 minutes longer.

YIELDS 1 9-INCH PIE

Apple Custard Pie

Peel sour apples and stew until soft and not much water left in them; then rub them through a colander. Beat 3 eggs for each pie to be baked, and put in at the rate of 1 cup butter and 1 of sugar for three pies; season with nutmeg.

Bake as pumpkin pies, which they resemble in appearance; and between them and apple pies in taste; very nice indeed.

From *Dr. Chase's Recipes,*
by A. W. Chase,
Ann Arbor, Michigan, 1874

Filling:
8   *large apples, peeled, cored, and sliced thick*
½   *cup water*
1   *cup granulated sugar*
4   *eggs, lightly beaten*
1   *teaspoon salt*
2   *teaspoons ground cinnamon*
1   *teaspoon ground ginger*
2   *tablespoons applejack or brandy*
1   *cup heavy cream*
2   *cups milk*

2   *recipes Pie Pastry for a one-crust pie (see page 232)*

Combine apples with water in a saucepan, cover, and bring to a boil. Reduce heat and simmer for about 20 minutes. Remove lid and, over gentle heat, boil off as much water as possible, taking care not to scorch the apples.

Preheat oven to 350 degrees F.

Work apples through a strainer with the back of a spoon and measure out 3 cups. Combine this in a large mixing bowl with sugar, beaten eggs, salt, cinnamon, ginger, applejack, heavy cream, and milk. Beat together well.

Line 2 9-inch pie pans with pastry dough. Pour the filling mixture equally into the pie shells.

Bake in a preheated oven for 45 minutes to 1 hour, or until custard is set. Test by sticking the blade of a knife into custard; if blade comes out clean, pies are done.

YIELDS 2 9-INCH PIES

Taffety-Tart

> *First, wet your paste with Butter, and cold water, roul it very thin, then lay Apples in lays, and between each lay of Apples strew some fine Sugar, and some Lemon-peel cut very small; you may also put some Fennel-seed to them, let them bake an hour or more, then ice them with Rose-water, Sugar, and Butter beaten together, and wash them over with the same, strew more fine Sugar over them, and put into the Oven again; this done, you may serve them hot or cold.*

> From *The Gentlewomans Companion*,
> by Hannah Woolley, London, 1673

The name Taffety-Tart could have stemmed from the notion that the iced crust of this pie had the sheen of a piece of taffeta cloth. However, in Hanna Woolley's time *taffety* was an adjective that could be applied to anything dainty and elegant. Either explanation fits this pie.

Pie Pastry for a two-crust pie (see recipe on page 232)

Filling:

5 to 6	*large apples, peeled, cored, and sliced thin*
½	*cup granulated sugar*
	Peel of ¼ lemon, grated
1	*teaspoon fennel or anise seeds, crushed slightly with mortar and pestle or with the back of a spoon*
1 to 2	*tablespoons butter (optional)*

Icing:

    2   *tablespoons butter, softened*
   ½   *cup confectioners' sugar*
    1   *teaspoon rosewater*
        Granulated sugar for garnish

Preheat oven to 450 degrees F.

Line a 9-inch pie pan with approximately half the pastry dough.

Combine granulated sugar, grated lemon peel, and crushed fennel together in a small mixing bowl.

Fill pie pan with apples in two or three layers, sprinkling over each layer some of the sugar mixture. Dot with butter, if desired.

Cover pie with top portion of pastry dough, then seal, trim, and flute edges, slashing the top in several places to let steam escape. Bake at 450 degrees F. for 15 minutes, then reduce heat and bake at 350 degrees F. for 35 to 40 minutes more. Remove from oven and let pie cool on a rack.

To make icing, beat 2 tablespoons softened butter in a mixing bowl until fluffy. Add confectioners' sugar gradually, beating well after each addition. Beat in the rosewater. Chill.

When upper crust of pie is just warm, ice with the rosewater icing. Sprinkle a little granulated sugar over the icing for garnish.

YIELDS 1 9-INCH PIE

Cranberry and Apple Pie

Take half a pint of cranberries, pick them from the stems and throw them into a saucepan with half a pound of white sugar and a spoonful of water; let them come to a boil; then retire them to stand on the hob while you peel and cut up four large apples; put a rim of light paste around your dish, strew and bake for an hour.

From *Godey's Lady's Book*,
Philadelphia, October 1866

Filling:

1   *cup sugar*
2   *tablespoons flour*
½   *teaspoon ground cinnamon*
¼   *teaspoon ground cloves*
1   *tablespoon grated orange peel*
¼   *cup water*
1   *tablespoon butter*
2   *cups peeled, cored, and coarsely diced apples*
3   *cups fresh cranberries*

Pie Pastry for a two-crust pie (see recipe on page 232)

Preheat oven to 425 degrees F.

To make filling, combine sugar, flour, cinnamon, cloves, grated orange peel, and water in a large saucepan. Cook over a low heat, stirring, until sugar is dissolved. Add the butter, and when it is melted, add the apples and cranberries. Bring to a boil and cook until the cranberries burst, about 5 minutes. Let cool.

Line a 9-inch pie pan with approximately half the pastry dough. Pour in the filling and arrange strips of pastry, lattice fashion, over the fruit. Pinch and flute the edges of the pastry, and bake in preheated oven for about 40 minutes, until crust is golden.

YIELDS 1 9-INCH PIE

Beef Mincemeat Pie

Four pound boild beef, chopped fine, and salted; six pound of raw apple chopped also, one pound beef suet, one quart of Wine or rich sweet cyder, one ounce mace, and cinnamon, a nutmeg, two pounds raisins.

From *American Cookery*,
by Amelia Simmons,
Hartford, Connecticut, 1796

Filling:

  2 *cups chopped boiled beef*
  ½ *cup chopped suet*
  6 *cups peeled, cored, and chopped apples*
  1 *cup seedless raisins*
  1 *cup sweet cider*
  1 *teaspoon ground cinnamon*
  1 *teaspoon ground mace*
  ½ *teaspoon grated nutmeg*
  1 *cup sugar*
  4 *tablespoons brandy*

  2 *recipes Pie Pastry for a two-crust pie (page 232)*

To make filling, combine all ingredients, except sugar and brandy, in a covered saucepan. Simmer until apples are soft, about 40 minutes. Remove from heat and stir in sugar, then brandy.

Preheat oven to 400 degrees F.

Line 2 9-inch pie pans with bottom portions of pastry dough. Pour half the filling into each. Cover with top portions of pastry dough or strips of pastry laid lattice fashion. If one whole layer of dough is used to cover pie, be sure to slash top crust in several places to allow steam to escape. Bake in preheated oven until crust is browned, about 35 minutes.

Although this recipe yields enough filling for 2 pies, there is no need to bake both pies at once. The mixture will keep for days in the refrigerator.

YIELDS 2 9-INCH PIES

Meatless Mince Pie

*To twelve apples chopped fine, add six beaten eggs, and a
half pint of cream. Put in spice, sugar, raisins or currants just as
you would for meat mince pies.*

From *The Young Housekeeper's Friend*,
by Mrs. (Mary Hooker) Cornelius,
Boston, 1859

Since this is not a real mince pie, it must be baked differently. The
filling is actually a type of egg custard.

Pie Pastry for a two-crust pie (see recipe on page 232)

Filling:

4 to 5	*apples, peeled, cored, and sliced thin*
1	*cup seedless raisins*
½	*cup light cream*
3	*eggs, well beaten*
¾	*cup sugar*
	Peel of ½ orange, grated
½	*teaspoon ground cinnamon*
½	*teaspoon ground cloves*
2	*tablespoons brandy*

Preheat oven to 350 degrees F.

Line a 9-inch pie pan with approximately half the pastry dough.

Combine in a mixing bowl ingredients for filling and mix
thoroughly.

Pour filling into pie pan and cover with strips of pastry put on
lattice fashion.

Bake in preheated oven until custard sets, about 1 hour. Test
custard by sticking the blade of a knife into it; if blade comes out
clean, custard is done.

YIELDS 1 9-INCH PIE

Apple and Green Tomato Pie

Take six good-sized apples and six large tomatoes; peel, core, and cut up the apples; put them into a glazed saucepan; squeeze the pulp from the tomatoes; put it with the apples; add a quarter of a pound of white sugar, and stir it over the fire until the apples begin to feel tender; then put an edge of puff-paste round a tart-dish; lay in your fruit, stirring in a couple of table-spoons of rich cream as you do so. Cover it with crust; place it in a moderately brisk oven, and bake for twenty minutes.

From *Godey's Lady's Book*,
Philadelphia, August 1867

The green tomatoes in this recipe provide the kind of tartness one seldom encounters in a fruit pie. Sometimes the tomatoes, depending on the variety of tomato used, can even impart a touch of bitterness. A fine dessert warm or cold, but not meant for anyone with a highly developed sweet tooth.

Filling:
   4   *apples, peeled, cored, and chopped*
   4   *green tomatoes, coarsely chopped*
   ⅔   *cup sugar*
   ⅛   *teaspoon salt*
   1   *tablespoon cornstarch*
   ¼   *teaspoon grated nutmeg*

   Pie Pastry for a two-crust pie (recipe on page 232)
   2   *tablespoons butter*

Preheat oven to 450 degrees F.

Combine chopped apples and tomatoes in a mixing bowl. Sprinkle over them the sugar, salt, cornstarch, and nutmeg. Mix gently but well.

Line a 9-inch pie pan with approximately half the pastry dough. Fill with the apple-tomato mixture, dot with butter, and cover with top portion of pastry dough. Seal and flute edges of crust, slashing top in several places to let out steam.

Bake at 450 degrees F. for 15 minutes, then reduce heat to 300 degrees and bake for about 45 minutes more.

YIELDS 1 9-INCH PIE

Cider Cake

Cider Cake.—Flour, 6 cups; sugar, 3 cups; butter, 1 cup; 4 eggs; cider, 1 cup; saleratus, 1 tea-spoon; 1 grated nutmeg.

Beat the eggs, sugar, and butter together, and stir in the flour and nutmeg, dissolve the saleratus in the cider and stir into the mass and bake immediately in a quick oven.

From *Dr. Chase's Recipes*,
by A. W. Chase,
Ann Arbor, Michigan, 1874

The recipe that follows is half the size of the original, and the nutmeg has been reduced substantially, in keeping with modern tastes. Saleratus is another name for baking soda (sodium bicarbonate).

> 3   cups sifted flour
> ½   teaspoon baking soda
> ½   teaspoon grated nutmeg
> ½   cup butter, softened
> 1½   cups sugar
> 2   eggs, well beaten
> ½   cup cider

Preheat oven to 350 degrees F.

In a mixing bowl, sift together flour, baking soda, and nutmeg. Set aside.

In another, large mixing bowl, beat together well the butter and sugar. Add eggs and beat well. Then add flour mixture and cider alternately, beginning and ending with the flour.

Spoon batter into a well-buttered 9¼ x 5¼ x 2¾-inch loaf pan, and bake for 1 hour in a preheated oven, or until a toothpick inserted into the center of the cake comes out dry.

YIELDS 1 9¼ x 5¼ x 2¾-INCH CAKE

Apple Cake

Batter:

2	*tablespoons butter, softened*
¼	*cup sugar*
1	*egg, lightly beaten*
2	*cups flour*
1½	*teaspoons baking powder*
½	*teaspoon salt*
¾	*cup milk*

3 to 4	*apples, peeled, cored, and sliced thick*
1 to 2	*tablespoons butter*
¼	*cup sugar*
1	*teaspoon ground cinnamon*

Preheat oven to 375 degrees F.

In a mixing bowl, cream together 2 tablespoons butter and ¼ cup sugar. Add egg and beat well.

In another mixing bowl, sift together flour, baking powder, and salt.

Add flour mixture and milk to egg mixture alternately, in two or three portions, to form a stiff, doughlike batter.

Spread batter on the bottom of a well-buttered 9-inch cake pan (a spring-form pan is even better). Cover batter with apple slices and dot with a tablespoon or two of butter.

Combine ¼ cup sugar and cinnamon in a small mixing bowl, and sprinkle the mixture evenly over the apples.

Bake in preheated oven for 45 minutes, or until a toothpick inserted into cake comes out clean.   YIELDS 1 9-INCH CAKE

Applesauce Cake

One cup sugar, one-fourth cup butter, one cup apple sauce (unsweetened), one cup raisins, one egg, one teaspoon cinnamon, one-fourth teaspoon cloves, little nutmeg, one teaspoon vanilla, two cups flour, one level teaspoon soda and a little salt.

From *The State of Maine Cook Book*,
edited by Jane Armstrong Tucker,
Wiscasset, Maine, circa 1924

1¾   *cups flour, sifted*
1   *teaspoon baking soda*
½   *teaspoon salt*
1   *teaspoon ground cinnamon*
¼   *teaspoon ground cloves*
   Pinch of grated nutmeg
½   *cup shortening*
1   *cup sugar*
1   *egg, well beaten*
1   *teaspoon vanilla*
1   *cup Applesauce (see recipe on page 200), prepared without sugar*
1   *cup seedless raisins*

Preheat oven to 350 degrees F.

In a mixing bowl, sift together flour, baking soda, salt, cinnamon, cloves, and nutmeg three times. Set aside.

In another large mixing bowl, cream together shortening and sugar until fluffy. Add beaten egg and vanilla, and beat thoroughly. Add flour mixture and Applesauce alternately in small quantities, beating well after each addition. Stir in raisins.

Butter an 8-inch-square baking dish and line bottom with wax paper. Pour in batter.

Bake in preheated oven for 45 minutes. Test cake by sticking a toothpick into center; if toothpick comes out clean, cake is done. Cool before serving.   YIELDS 1 8-INCH-SQUARE CAKE

Dried-Apple Cake

One and one-half cups dried apples, to be soaked over night, in the morning chop them and simmer in ½ cup of molasses until they are well cooked. Stir to a cream. One-half cup butter and 1 cup sugar, add 2–3 cup of sour milk and 2 eggs beaten well. Two teaspoons of cinnamon, 1 teaspoon of cloves, 1 teaspoon of nutmeg, 1 teaspoon soda dissolved in a little water. After mixing well together stir in the apples and 2 cups of flour, add 1 cup raisins, ½ cup currants, 1 ounce sliced citron. Bake in a moderate oven one hour.

From *The Palisades Cook Book*,
by the Ladies Aid Society of
the Tenafly, New Jersey,
Presbyterian Church, 1910

1½	cups dried apple slices*
½	cup molasses
½	cup butter, softened
1	cup sugar
2	eggs, well beaten
⅔	cup sour milk or buttermilk**
1	cup seedless raisins
½	cup dried currants
2	tablespoons diced candied citron
2	cups flour
1	teaspoon baking soda
½	teaspoon salt
2	teaspoons ground cinnamon
1	teaspoon grated nutmeg
1	teaspoon ground cloves

* If dried apples are unavailable, substitute 1⅓ cups thick, unsweetened applesauce for the dried apple slices and the molasses and replace the 1 cup granulated sugar with 1 cup tightly packed, dark brown sugar.
** If neither sour milk nor buttermilk are available, replace with sweet milk and substitute 2 teaspoons baking powder for the baking soda.

Cover dried apple slices with water and soak overnight.

The next day, drain, chop, and cook the apples with the molasses in a saucepan until they are tender, about ½ hour. Beat until smooth and set aside.

Preheat oven to 350 degrees F.

In a mixing bowl, combine butter and sugar as best you can, then beat in the eggs. Add sour milk and beat well. Stir in the apple–molasses mixture.

Combine the raisins, currants, and citron in a mixing bowl, and sprinkle a bit of flour over all to keep fruit from sticking together.

In another bowl, sift the remaining flour with the baking soda, salt, cinnamon, nutmeg, and cloves.

Stir the flour mixture gradually into the liquid until smooth. Stir in the fruits.

Pour batter into a buttered 9-inch tube pan, and bake in pre-heated oven for about 1 hour. Test cake by sticking a toothpick into it; if toothpick comes out dry, cake is done.

YIELDS 1 9-INCH CAKE

13

Puddings and Other Desserts

EVE'S PUDDING
INDIAN AND APPLE PUDDING
BLACKBERRY APPLE PUDDING
APPLE AND PLUM PUDDING
APPLE CREAM
APPLE COBBLER
APPLE PAN DOWDY
APPLE SLUMP (APPLE POT PIE)
BROWN BETTY
APPLE FRAISE
APPLE CHARLOTTE
APPLE SOUFFLÉ
APPLE SNOW
BLACK CAPS (BAKED APPLES)

Eve's Pudding

Grate three-fourths of a pound of stale bread, and mix it with three-fourths of a pound of fine suet, the same quantity of chopped apples and dried currants, five eggs, and the rind of a lemon. Put it in a mould and boil it three hours. Serve it with sweet sauce.

From *The Dollar Monthly Magazine*,
Boston, May 1864

APPLE FRITTERS

Etching by J. P. LeBas. Cookouts in eighteenth-century France, as this illustration shows, were as elaborate as anything dreamed up in twentieth-century America and really quite similar.

The recipe that follows has been reduced to a more reasonable size, and the flavor has been lightened by substituting butter for the suet, which many people today find a bit too heavy on the tongue.

- 1¼   *cups bread crumbs*
- 1   *apple, peeled, cored, and chopped*
- ⅓   *cup seedless raisins*
-    *Peel of ¼ lemon, grated or chopped fine*
- ⅓   *cup sugar (optional)*
- ¼   *teaspoon grated nutmeg (optional)*
- 1   *teaspoon ground cinnamon (optional)*
- 2   *eggs, lightly beaten*
- 1   *teaspoon baking powder*
- ¼   *cup melted butter*
- 2   *tablespoons (approximately) milk*
- ½   *cup Hard Sauce (see recipe on page 255)*

Combine all the ingredients, except the milk and the Hard Sauce, in a mixing bowl in the order listed. Mix all together well, then add just enough milk so that the mixture is stiff but manageable.

Spoon into a well-buttered, 1-quart pudding mold or into a large buttered tin can. If the mold has no cover, fashion one with aluminum foil.

The pudding can be steamed in either of two ways: If the mold will fit inside a pressure cooker, put it there, and fill the pot with enough water to come halfway up the sides of the mold. Steam the pudding for ½ hour without pressure, then at a pressure of 15 pounds per square inch for ½ hour more.

Alternatively, the mold can be placed in a large pot that has a cover. Fill the pot with water so that the water comes halfway up the sides of the mold. Cover the pot and boil the pudding gently for about 3 hours, being careful to add more water to the pot should any boil away.

Slice pudding, put into individual dessert dishes, and place a dollop of Hard Sauce on each serving. Serve hot.

SERVES 6

HARD SAUCE

⅓   *cup butter, softened*
¾   *cup confectioners' sugar*
2   *tablespoons brandy*

Cream butter and sugar together in a mixing bowl. Gradually add brandy, and beat just until sauce is smooth. Chill for a few minutes.

YIELDS ½ CUP

Indian and Apple Pudding

*One cupful of Indian meal, one cupful of molasses, two
quarts of milk, two teaspoonfuls of salt, three tablespoonfuls
of butter, or one of finely chopped suet; one quart of pared and
quartered apples (sweet are best, but sour will do), half a teaspoonful of ginger, half a teaspoonful of grated nutmeg. Put the
milk on in the double boiler. When it boils, pour it gradually
on the meal. Pour into the boiler again and cook half an hour,
stirring often. Add the molasses, butter, seasoning and apples, and
bake slowly three hours. Make half the rule if the family is small.*

From *Miss Parloa's New Cook Book*,
by Maria Parloa, Boston, 1880

6	*cups milk*
½	*cup yellow corn meal (Indian meal)*
⅔	*cup molasses*
2	*tablespoons butter*
1	*teaspoon salt*
½	*teaspoon ground ginger*
¼	*teaspoon grated nutmeg*
3	*apples, peeled, cored, and sliced thin*
	Light cream (optional)

Preheat oven to 275 degrees F.

Scald 4 cups of the milk in the top of a double boiler over direct
heat, then place over boiling water.

Gradually add corn meal to the hot milk and cook, stirring constantly, until mixture is thick.

Add molasses, butter, salt, ginger, and nutmeg. Remove from heat
and stir in remaining 2 cups of cold milk.

Grease a 3-quart baking dish and put in a layer of the corn-meal
mixture. Spread apple slices over that, then put in more corn-meal
mixture and apples, alternating until you have 2 or 3 complete layers.

Bake 3 hours in preheated oven.

Slice pudding, put into individual serving dishes while hot, and top each portion of pudding with light cream, if desired.

SERVES 12

Blackberry Apple Pudding

Take a quart of blackberries, six large apples, peeled and sliced in thin pieces, half a pound of sugar, and three or four slices of lemon-peel; make a light paste, line a deep dish, and fill it with the above ingredients, and let the pudding boil steadily for one hour. A little grated nutmeg, with a small cup of sweet cream upon it, will render it a most delicious viand.

From *Godey's Lady's Book,*
Philadelphia, August 1866

Dough:

    1   *cup sifted flour*
    ¼   *teaspoon salt*
    ⅓   *cup vegetable shortening*
2 to 3   *tablespoons ice water*

    2   *cups fresh blackberries (1 pint basket)*
    3   *large apples, peeled, cored, and sliced thin*
¼ to ½   *cup sugar*
         Peel of ½ lemon, grated
         Light cream

To prepare the dough, sift together flour and salt into a mixing bowl. Cut in the shortening with a pastry blender or a pair of knives until it is cut into bits the size of small peas. Add ice water, a tablespoonful at a time, and toss dough with a pair of forks after each addition. Continue to add ice water until a bit of the dough will form a ball when lightly squeezed.

Roll out the dough on a well-floured surface, and line a 1-quart mixing bowl (glass, ceramic, or metal) with the pastry.

In another mixing bowl, lightly toss together the blackberries and the apple slices.

Combine the sugar and the grated lemon peel in a separate bowl.

Put a layer of fruit into the pastry-lined bowl and sprinkle some of the sugar mixture over the fruit. Continue in this fashion, following each layer of fruit with a sprinkling of sugar until bowl is filled with about 3 or 4 complete layers.

Cover bowl with a piece of aluminum foil and put it inside a large covered pot. Add water to the pot until the liquid comes halfway up the sides of the bowl. Put the covered pot over high heat, bringing the water to a boil. Reduce heat, and steam the pudding for about 1 hour.

Slice hot pudding, put into individual serving bowls, and top each serving with light cream.

SERVES 6 TO 8

Apple and Plum Pudding

¾ pound fine tart apples, pared and chopped; ¾ pound sugar; ¾ pound flour; ½ pound beef suet, rubbed fine; ¾ pound raisins, seeded and chopped; 6 eggs; 1 teaspoonful nutmeg and the same powdered cloves; 1 teaspoonful salt; ½ glass brown Sherry and the same of brandy.

From *Common Sense in the Household*,
by Marion Harland, New York, 1884

This recipe will feed a small army, but, fortunately, it can be easily halved. In any case, leftover pudding will keep for days and can be reheated by steaming it on a rack over boiling water inside a covered pot. As with the traditional British plum pudding, this recipe contains nary a plum. For some unexplained reason, raisins in pudding are called plums and have been for centuries.

 3   eggs, separated
 1   cup sugar
½   cup beef suet, chopped fine
 2   cups peeled, cored, and chopped apples
½   teaspoon grated nutmeg
½   teaspoon ground cloves
½   teaspoon salt
 1   cup chopped raisins
1½ cups flour
 2   tablespoons sherry
 2   tablespoons brandy
     Recipe Hard Sauce (see recipe on page 255)

Preheat oven to 325 degrees F.

Beat egg yolks in a bowl and combine with sugar. Add suet, chopped apples, nutmeg, cloves, and salt.

Dredge raisins in flour, and add both the raisins and the remaining flour to the mixture. Add sherry and brandy.

Beat egg whites until they are very stiff and add to the mixture. Combine all well.

Grease an 8-inch-square baking pan and pour in the mixture.

Bake in a preheated oven for 1½ hours. Before taking the pudding from the oven, test it by sticking a toothpick into the center; if toothpick comes out clean, the pudding is done.

Serve hot with Hard Sauce.   SERVES 10

Apple Cream

Take a dozen Pippins, or more, pare, slice, or quarter them, put them into a Skillet, with some Claret wine, and a race of Ginger sliced thin, a little Lemon-peel cut small, and some Sugar; let all these stew together till they be soft, then take them off the fire, and put them into a dish, and when they be cold, take a quart of boil'd Cream, with a little Nutmeg, and put in of the Apple as much as will thicken it, and so serve it up.

From *The Gentlewomans Companion*,
by Hannah Wooley, London, 1673

This dish resembles the cold fruit soups popular as desserts in Scandinavia and Central Europe. The amounts specified in the original recipe are for a large household; the modern version is easily halved.

6   *large apples, peeled, cored, and quartered*
1   *cup dry red wine*
1   *cup sugar*
1   *1-inch-long piece gingerroot or 1 teaspoon ground ginger*
    Peel of ¼ lemon, cut into julienne strips
1   *cup heavy cream*
    Freshly grated nutmeg for garnish

Combine apple quarters in a kettle with wine, sugar, gingerroot, and lemon peel. Stew for 1 hour. Allow to cool.

Add heavy cream and mix well.

Pour into individual serving dishes and garnish with grated nutmeg. Serve well chilled.

SERVES 6

Apple Cobbler

6   *apples, peeled, cored, and sliced thick*
½   *cup water*
1   *cup sugar*

Dough:
1¾   *cups flour, sifted*
2½   *teaspoons baking powder*
½   *teaspoon salt*
1   *tablespoon sugar*
¼   *cup butter, chilled*
¾   *cup milk*
1   *teaspoon ground cinnamon*
½   *cup Hard Sauce (see recipe on page 255)*

Combine sliced apples and water in a covered saucepan. Bring to a boil, reduce heat, and simmer until apples are just tender, about 20 minutes. Add 1 cup of sugar and simmer until sugar is dissolved, about 5 minutes more.

Preheat oven to 425 degrees F.

While apples are cooking, sift together flour, baking powder, salt, and 1 tablespoon of sugar into a mixing bowl. Cut in chilled butter with a pair of knives or a pastry blender, until the mixture has the consistency of coarse meal. Stir in milk.

Turn out dough onto a well-floured surface. Knead for about 30 seconds.

Butter an 8-inch-square baking dish. Add the apples while they are still boiling hot. Sprinkle cinnamon over apples. Form dough to roughly the size of the baking dish and cover the apples with the dough. Bake in a preheated oven for ½ hour.

Turn cobbler out onto a serving plate, upside down, with the apples on top. Serve with Hard Sauce.   SERVES 10

Apple Pan Dowdy

Take a deep, brown baking-pan; butter it; fill it with apples, peeled, cut in quarters, and cored; add a large spoonful of cinnamon, two teacups of brown sugar, one teacupful of good cider, if you have it; if not, a little water; cover with a common pie-crust; bake about four hours; then break the crust into the pan with the apples and juice. To be eaten with sugar and cream.

From *The Dollar Monthly Magazine*,
Boston, November 1864

6	*large apples, peeled, cored, and thickly sliced*
½	*cup cider (or water)*
¾	*cup firmly packed dark brown sugar*
¾	*teaspoon salt*
1	*teaspoon ground cinnamon*
7	*tablespoons butter*

Crust:

1   *cup flour*
2   *teaspoons baking powder*
¾ *cup milk*
    Light cream

Preheat oven to 400 degrees F.

Butter an 8-inch-square baking dish and cover bottom with apple slices. Pour cider over apples.

In a mixing bowl, combine brown sugar, ¼ teaspoon salt, and cinnamon, and sprinkle this evenly over the apples. Dot with 4 table-spoons butter. Set aside.

To prepare crust, in a bowl sift together flour, ½ teaspoon salt, and baking powder. Cut in the 3 tablespoons butter, using either a pastry blender or a pair of knives, until butter is cut into bits the size of small peas. Stir in the milk. Spread this batter over the apple slices in the baking dish.

Bake in a preheated oven for about 40 minutes.

When serving, break up the crust and spoon it and the apples into individual serving dishes. Serve warm, topped with light cream.

SERVES 10

Apple Slump (Apple Pot Pie)

1 pint flour; 1½ teaspoons baking powder; 1 coffee cup sweet milk; Pinch of salt. Take nice juicy apples, pare and cut them into slices, put in a stew pan, partly cover with cold water. Let them cook until soft, drop crust by spoonfuls on top of apples. Cover tightly and let steam till done. Serve with cream an sugar.

From *The Palisades Cook Book,*
by the Ladies' Aid Society of
the Tenafly, New Jersey,
Presbyterian Church, 1910

Frying fritters in the seventeenth century brought the cook to the fire's edge. A long-handled skillet was a must, and here two of the women are shown shielding their faces from the intense heat—one using a fanlike device, the other with a handkerchief. From the *Art Journal*, London, 1854.

6	*large apples, peeled, cored, and sliced thick*
1¼	*cups sugar*
½	*cup water*
1	*teaspoon ground cinnamon*
1½	*cups flour, sifted*
1½	*teaspoons baking powder*
	Pinch of salt
2	*tablespoons butter, softened*
¾	*cup milk*

In a large saucepan, combine apple slices, 1 cup sugar, water, and cinnamon. Cover, place over high heat, and bring to a boil. Reduce heat and simmer for 10 minutes.

Meanwhile, sift together flour, baking powder, and salt into a mixing bowl. Set aside.

Cream butter and ¼ cup sugar together in a large bowl. Add flour mixture and blend well. Gradually add milk, and mix all well.

When apples have simmered for 10 minutes, remove cover from saucepan and drop in spoonfuls of the dough. Cover pot tightly and simmer until dumplings are done, 20 to 25 minutes more. Serve in individual serving dishes with some apples on the bottom, a dumpling on top, and some syrup from the pot spooned over all.

SERVES 6

Brown Betty

1   *cup bread crumbs*
2   *cups chopped apples, tart*
½   *cup sugar*
1   *teaspoon {ground} cinnamon*
2   *tablespoons butter cut into small bits*

Butter a deep dish, and put a layer of the chopped apple at the bottom; sprinkle with sugar, a few bits of butter, and cinnamon; cover with bread-crumbs; then more apple. Proceed in this order until the dish is full, having a layer of crumbs at top. Cover closely, and steam three-quarters of an hour in a moderate {350 degrees F.} oven; then uncover and brown quickly.

Eat warm with sugar and cream, or {a} sweet sauce.

This is a homely but very good pudding, especially for the children's table. Serve in the dish in which it is baked.

From *Common Sense in the Household*,
by Marion Harland, New York, 1884

The traditional accompaniment for Brown Betty is Hard Sauce (see recipe on page 255).

SERVES 4

Apple Fraise

Pare six large apples, take out the cores, cut them in slices, and fry them on both sides with butter; put them on a sieve to drain, mix half a pint of milk to a batter, not too stiff, put in a little lemon peel shred fine, a little beated cinnamon, put some butter into a frying pan, and make it hot; put in half the batter and lay the apples on it, let it fry a little to set it, then put the other batter over, fry it on one side, then turn it and fry the other side brown; put it into a dish, strew powder sugar over it, and squeeze over it also the juice of a Seville orange.

From *The New Art of Cookery*,
by Richard Briggs,
Philadelphia, 1792

Batter:

1½	*cups flour*
1	*teaspoon baking powder*
1	*cup milk*
2	*eggs, lightly beaten*
1	*teaspoon finely shredded lemon peel*
1	*teaspoon ground cinnamon*
6	*large apples, peeled, cored, and sliced thin*
3	*tablespoons butter*
	Confectioner's sugar for garnish
	Juice of 1 orange

To prepare batter, sift flour with baking powder into a mixing bowl. Combine this with milk, eggs, shredded lemon peel and cinnamon.

Fry apple slices in 2 tablespoons of butter until they are golden brown. Remove from skillet and drain on paper towels.

Heat the remaining 1 tablespoon of butter in a large skillet and add half the batter. Lay the fried apple slices over this in a single even layer. Let it "set" for about 3 or 4 minutes, then add the remaining batter. When the underside of the "pancake" is golden, turn it over and brown the other side.

Transfer to a heated platter and sprinkle with confectioner's sugar and orange juice.

SERVES 6 TO 8

Apple Charlotte

Take any number of rennet apples you may desire to use; peel them, cut them into quarters, and take out the core. Cut the quarters into slices, and let them cook over a brisk fire, with butter, sugar, and powdered cinnamon, until they are en mar-malade. *Cut thin slices of crust of bread, dip them in butter, and with them line the sides and bottom of a tin shape. Fill the middle of the shape with alternate layers of the apple and any preserve you may choose, and cover it with more thin slices of bread. Then place the shape in an oven, or before the fire, until the outside is a fine brown, and turn it out upon a dish, and serve it either hot or cold.*

From *Peterson's Magazine,*
Philadelphia, January 1865

This classic apple dessert predates the more famous cream-filled Charlotte Russe by many years, and while the Charlotte Russe was named after the Apple Charlotte, this apple dessert apparently was named after a character in a novel. The novel was Goethe's *The Sorrows of Werther*, and the hero, Werther, sees his beloved Charlotte for the first time through a window, as she is slicing bread and buttering it for her many little brothers and sisters. The buttered bread encasing the Apple Charlotte apparently is the link.

10	*large apples, quartered*
¼	*cup water*
11	*tablespoons butter*
½	*teaspoon ground cinnamon*
¼	*teaspoon grated nutmeg*
	Juice and grated peel of ½ lemon
½	*cup sugar*
10 to 12	*slices "homestyle" white bread, with crusts removed*
	Whipped cream (optional)

Combine apples and water in a heavy saucepan. Cover, place over high heat, and bring apples to a boil. Reduce heat and boil gently for 20 to 30 minutes, until apples are tender.

Run the cooked apples through a food mill or force them through a strainer, using the back of a wooden spoon. Put resulting applesauce back in saucepan and boil over high heat for 20 minutes, stirring frequently, to reduce to a thick purée.

Remove from heat and stir in the 3 tablespoons of the butter, cinnamon, nutmeg, lemon juice and grated peel, and sugar. Set aside.

Preheat oven to 375 degrees F.

Line a bread pan or similar form of 1-quart capacity with bread slices. Cut bread so slices cover bottom and sides of pan as tightly as possible. Reserve excess pieces of bread.

Melt the remaining 8 tablespoons (¼ pound) of butter in a small skillet, quickly dip each slice of bread into the butter, lightly coating both sides, and return it to its proper place in the mold.

Fill bread-lined mold with apple mixture and cover the apples with the bits and pieces of bread trimmed off the other slices. Dip bread pieces first in butter, then fit them together over the apples. (This will become the bottom of the Charlotte when it is turned out of the mold, so there is no need to be too fancy.)

Bake in a preheated oven for about 45 minutes, until the top crust is deeply browned. Take Charlotte out of the oven and let it cool for ½ hour. Then turn it over onto a flat plate, without removing the mold, and let it cool for at least ½ hour more.

Carefully remove mold, top each portion with whipped cream, if desired, and serve.

SERVES 12

Apple Soufflé

Stew the apples with a little lemon peel; sweeten them, then lay them pretty high round the inside of a dish. Make a custard of the yelks of two eggs, a little cinnamon, sugar, and milk. Let it thicken over a slow fire, but do not boil; when ready, pour it in the inside of the apple. Beat the whites of the eggs to a strong froth, and cover the whole. Throw a good deal of pounded sugar over it, and brown to a fine brown.

From *Godey's Lady's Book*,
Philadelphia, September 1866

This is a stewed apple and custard dessert, not a modern soufflé.

8	*large apples, peeled, cored, and sliced thick*
	Peel of ½ lemon
¼	*cup water*
¾	*cup sugar*

Custard:

1	*cup milk*
2	*egg yolks*
	Pinch of salt
2	*tablespoons sugar*
¼	*teaspoon ground cinnamon*

Meringue:

2	*egg whites*
4	*tablespoons sugar*
	Few grains of salt
¼	*teaspoon lemon extract (optional)*

Combine apple slices, lemon peel, and water in a saucepan. Bring to a boil, cover saucepan, and reduce heat. Simmer apples until they are tender, about 20 to 25 minutes.

Remove lemon peel, mash apples, and stir in ¾ cup sugar. Cook apples uncovered for 10 minutes more to dissolve sugar and to thicken mixture slightly.

Pour apples into an oven-proof 3-quart dish and set aside to cool.

To make custard, scald milk in the top of a double boiler. In a mixing bowl, lightly beat together 2 eggs yolks, a pinch of salt, 2 tablespoons sugar, and cinnamon. Gradually add the scalded milk, stirring constantly, and return mixture to the top of the double boiler. Cook mixture over hot water, stirring constantly, until mixture coats a spoon. Remove custard from heat.

Make a little hollow in the middle of the cooled applesauce in the baking dish, and pour the custard into this hollow.

Preheat oven to 350 degrees F.

To make meringue, beat 2 egg whites in a bowl until frothy. Add a few grains of salt, then add 4 tablespoons sugar gradually, beating all the while, and continue beating until egg whites are stiff. Add lemon extract. Spread meringue over apples and custard.

Bake in a preheated oven until meringue is lightly browned, about 10 to 15 minutes.

Serve warm.

SERVES 10

Apple Snow

Put twelve large apples, without paring, into cold water enough to stew them. Boil them slowly; when they are very soft strain them through a sieve; beat the whites of twelve eggs to a stiff froth, then add to them half a pound of fine white sugar, and when these are well mixed, add the apple, and beat all together, until white as snow. Then lay in the center of a deep dish, heap it as high as you can, and pour around it a nice boiled custard made of a quart of milk and eight of the yolks of the eggs.

From *The Young Housekeeper's Friend*,
by Mrs. (Mary Hooker) Cornelius,
Boston, 1859

This recipe is easily halved.

6   large apples
6   egg whites
½   cup sugar

Custard:
4   egg yolks
¼   teaspoon salt
½   cup sugar
2   cups milk
1   teaspoon vanilla extract

Place apples (it is not necessary to peel them) into enough water to cover, and stew slowly for about 1 hour. When they are very soft, strain through a sieve.

Beat the 6 egg whites in a bowl until they are stiff. Add ½ cup sugar, and when this is well combined, add strained apples. Beat all together until it is "white as snow." Put mixture in refrigerator to chill while preparing custard.

To prepare custard, lightly beat 4 egg yolks in the top of a double boiler. Add ¼ teaspoon salt and ½ cup sugar. Scald milk slowly in a saucepan and add it to the egg-yolk mixture. Cook in top of a double boiler over gently boiling water until mixture is able to coat a spoon. When the custard is ready, remove from heat and stir in vanilla extract. Let custard cool for a few minutes before serving.

To serve, lay the apple mixture in the center of a deep dish, heaping it as high as you can, and pour the custard around it. Serve immediately.

SERVES 6 TO 8

Black Caps
(Baked Apples)

Take out the cores, and cut into halves twelve large apples. Place them on a tin patty-pan as close as they can lie, with the flat side downwards. Squeeze a lemon into two spoonsful of orange-

flower water, and pour it over them. Shred some lemon-peel fine, and throw it over them, and grate fine sugar over all. Set them in a quick oven, and half an hour will do them. When you send them to table, strew fine sugar over the dish.

From *Modern Domestic Cookery*,
by W. A. Henderson,
New York, 1857

W. A. Henderson left the mystery of how this dish got its name to posterity.

12 *large apples, halved and cored*
    Juice of 1 lemon
 2 *tablespoons orange-flower water*
    Peel of ½ lemon
 1 *cup sugar*
    Sugar for garnish
    Light cream (optional)

Preheat oven to 350 degrees F.

Choose a baking dish that will accommodate all the apple halves, flat side down, with as little room to spare as possible. Butter the dish well, then fill it with the apple halves.

Combine lemon juice with orange-flower water. Sprinkle over the apples. Grate lemon peel directly over apples, then sprinkle 1 cup sugar over all.

Bake in preheated oven until apples are tender, about 30 to 40 minutes.

Arrange apples, flat side up, on a serving dish. Pour any syrup remaining in the baking dish over them. Garnish with a little more sugar before serving. Serve warm or cold as a dessert, topping individual portions with a little light cream, if desired. These apples can also be served hot as a relish with roasted fatty meats, such as pork, goose, or duck.

SERVES 12

Bibliography

Apicius, Coelius. *De Re Coquinaria.* Heidelbergae: C. Winter, 1867.

Baring-Gould, S. *Curious Myths of the Middle Ages.* London: Longmans, Green & Co., 1897.

Beach, S. A. *The Apples of New York State.* Albany: J. B. Lyon Co., 1905.

Beeton, Mrs. (Isabella). *The Book of Household Management.* London: Ward, Lock and Co., 1861.

Black, William George. *Folk Medicine.* London: Folk-Lore Society, 1883.

Briggs, Richard. *The New Art of Cookery.* Philadelphia: W. Spotswood et al, 1792.

Campbell, Joseph. *The Masks of God: Occidental Mythology.* New York: The Viking Press, 1964.

Cannell, Margaret. "Signs, Omens and Portents in Nebraska Folklore." In *University of Nebraska Studies in Language, Literature and Criticism,* no. 13. Lincoln, Nebr.: University of Nebraska Press, 1933.

Carlson, R. F., et al. *North American Apples: Varieties, Rootstocks, Outlook.* East Lansing, Mich.: Michigan State University Press, 1970.

Cato, Marcus Porcius. *De Agricultura.* Translated by Ernest Brehaut. New York: Columbia University Press, 1933.

Chase, A. W. *Dr. Chase's Recipes.* Ann Arbor, Mich.: R. A. Beal, 1874.

Child, Mrs. (Lydia Maria). *The American Frugal Housewife.* Boston: Carter, Hendee, and Co., 1833.

Christ, E. G. *How to Prune Young and Bearing Apple Trees* (Extension Bulletin 377-A). New Brunswick, N.J.: Cooperative Extension Service of Rutgers University, undated.

Cornelius, Mrs. (Mary Hooker). *The Young Housekeeper's Friend*. Boston: Brown, Taggard and Chase, 1859.

Coxe, William. *A View of the Cultivation of Fruit Trees in the United States, and of the Management of Orchards and Cider*. Philadelphia: M. Carey & Son, 1817.

Dafydd ap Gwilym. *Fifty Poems*. Translated by Idris Bell and David Bell. London: The Honourable Society of Cymmrodorian, 1942.

Dawson, Thomas. *The Good Husvvifes Ievvell*. London: John Wolf for Edward White, 1587.

Digby, Kenelm. *Choice and Experimented Receipts in Physicke and Chirurgery*. London: Printed for the author, 1668.

Dioscorides. *The Greek Herbal of Dioscorides, Illustrated by a Byzantine A.D. 512, Englished by John Goodyear A.D. 1655, Edited and First Printed A.D. 1933 by Robert T. Gunther*. New York: Oxford University Press, 1934.

Downing, A. J. *The Fruits and Fruit Trees of America*. Revised and corrected by Charles Downing. New York: John Wiley, 1859.

Earle, Alicia Morse. *Old Time Gardens*. New York: The Macmillan Co., 1901.

Fairbrother, Nan. *Men and Gardens*. New York: Alfred A. Knopf, 1956.

Farmer, Fannie Merritt. *The Boston Cooking-School Cook Book*. Boston: 1896.

Foster, B. O. "Notes on the Symbolism of the Apple in Classical Antiquity." In *Harvard Studies in Classical Philology* 10 (1899).

Frazer, James George. *The Golden Bough*. 3rd ed. 12 vols. London: Macmillan and Company, 1911–1926.

Frothingham, A. L. "Medusa, Apollo, and the Great Mother." In *American Journal of Archaeology*, Second Series, 15 (1911).

Gerarde, John. *The Herball, or Generall Historie of Plantes*. London: John Norton, 1597.

Glasse, Hannah. *The Art of Cookery Made Plain and Easy*. London: A. Millar, et al, 1767.

Graves, Robert. *The White Goddess*. New York: Creative Age Press, 1948.

———. *The Greek Myths*. 2 vols. Baltimore, Md.: Penguin Books, Inc., 1955.

Gray, Louis Herbert, ed. *The Mythology of All Races in Thirteen Volumes*. Vols. 1–3. Boston: Marshall Jones Co., varying dates.

Grimm, Jakob and Karl. *Kinder- und Haus-Maerchen*. 1812. Reprint. Hamburg: Fischer Buecherei, 1962.

Guest, Lady Charlotte. *The Mabinogion*. London: Longmans, Green, and Co., 1849.

Halliwell, J. O., ed. *Early English Miscellanies in Prose and Verse*. London: Warton Club, 1855.

Harland, Marion. *Common Sense in the Household*. New York: Charles Scribner's Sons, 1884.

Hedrick, U. P. *Fruits for the Home Garden*. New York: Oxford University Press, 1944.

————. *A History of Horticulture in America to 1860*. New York: Oxford University Press, 1950.

Henderson, W. A. *Modern Domestic Cookery*. New York: Leavitt & Allen, 1857.

Joyce, P. W. *A Social History of Ancient Ireland*. London: Longmans, Green, & Co., 1920.

Ladies' Aid Society of the Tenafly, New Jersey, Presbyterian Church. *The Palisades Cook Book*. Tenafly, N.J.: Englewood Press, 1910.

Le Menagier de Paris. Translated by Elizabeth Power as *The Goodman of Paris*. London: George Routledge & Sons, 1928.

Leslie, Miss (G. C.). *Seventy-five Receipts for Pastry, Cakes, and Sweetmeats*. Boston: Monroe and Francis, 1832.

McCartney, E. S. "How the Apple Became a Symbol of Love." In *Transactions and Proceedings of the American Philological Association*, 56 (1925).

[Markham, Gervase]. *The English Hus-wife*. London: John Beale for Roger Jackson, 1615.

Neumann, Erich. *The Great Mother*. Princeton, N.J.: Princeton University Press, 1925.

Nicholson, Katherine Stanley. *Historic American Trees*. New York: Fry Publishing Co., 1922.

O'Rahilly, Thomas F. *Early Irish History and Mythology*. Dublin: The Dublin Institute for Advanced Studies, 1946.

Pareti, Luigi. *History of Mankind. Vol. 2: The Ancient World*. New York: Harper & Row, 1965.

Parkinson, John. *Paradisi in Sole Paradisus Terrestris*. London: 1629.

Parloa, Maria. *Miss Parloa's New Cook Book and Marketing Guide*. Boston: Estes and Lauriat, 1880.

Pegge, Samuel, ed. *The Forme of Cury*. London: J. Nichols, 1780.

Pliny [Caius Plinius Secundus]. *Historia Naturalis*. Translated by Harris Rackham, et al. Cambridge: Harvard University Press, 1938–1962.

Puckett, Newbell Niles. *Folk Beliefs of the Southern Negro*. Chapel Hill, N.C.: University of North Carolina Press, 1926.

Pullar, Philippa. *Consuming Passions*, Boston: Little, Brown and Co., 1970.

Rhys, John. *Celtic Folklore—Welsh and Manx*. Vols. 1 and 2. Oxford, Eng.: Clarendon Press, 1901.

Rice, Rosella. "Recollections." In *History of the Ashland County Pioneer Historical Society*. Ashland, Ohio, 1888.

Simmons, Amelia. *American Cookery*. Hartford, Conn.: Hudson & Goodwin, 1796.

Skinner, Charles M. *Myths and Legends of Flowers, Trees, Fruits, and Plants*. Philadelphia: J. P. Lippincott Co., 1911.

Thomas, Daniel Lindsay, and Lucy Blayney. *Kentucky Superstitions*. Princeton, N.J.: Princeton University Press, 1920.

Tucker, Jane Armstrong, ed. *The State of Maine Cook Book*. Wiscasset, Maine: The Democratic Women of Maine, circa 1924.

Tyree, Marion Cabell. *Housekeeping in Old Virginia*. Louisville, Ky.: John P. Morton and Co., 1879.

W.M. *The Queens Closet opened*. London: Nathaniel Brook, 1655.

————. *The Compleat Cook*. London: Nathaniel Brook, 1655.

Whitney, A. W., and Bullock, C. C. *Folk Lore from Maryland*. New York: American Folk Lore Society, 1925.

Wood-Martin, W. G. *Traces of the Elder Faith in Ireland*. London: Longmans, Green, & Co., 1902.

Woolley, Hannah. *The Gentlewomans Companion; or, a Guide to the Female Sex*. London: A. Maxwell for Dorman Newman, 1673.

Wright, Richardson. *The Story of Gardening*. New York: Dodd, Mead and Co., 1934.

INDEX